The
Triumph
of the Darwinian
Method

MICHAEL T. GHISELIN

The Triumph of the Darwinian Method

1969

UNIVERSITY OF CALIFORNIA PRESS

BERKELEY AND LOS ANGELES

University of California Press
Berkeley and Los Angeles, California
University of California Press, Ltd.
London, England
Copyright © 1969 by
The Regents of the University of California
Library of Congress Catalog Card Number: 69-15938
Printed in the United States of America

To Brewster Ghiselin

Acknowledgments

In preparing this work, I have been indeed fortunate in knowing scholars who gladly came to my assistance with their time, their erudition, and their good taste. This strikes me as a good sign; not only that many generous people are to be found in the scientific community, but that so many share my conviction that the Darwinian endeavor has profound meaning for all of us. And it seems that in spite of a trend toward narrow specialization, the various scientific disciplines may still find their proper bond of union in general ideas. That we continue to draw inspiration from the writings of biology's greatest generalist seems to justify a feeling that we have perpetuated the liberal tradition which he embodied so well.

I wish to express my deepest gratitude to those who have criticized parts of the manuscript: Melbourne R. Carriker, Philip J. Darlington, Richard J. Herrnstein, Ernst Mayr, Everett Mendelsohn, Frank Pitelka, Thomas J. M. Schopf, Mary P. Winsor and Victor A. Zullo. I am pleased to acknowledge the use of the libraries at the Marine Biological Laboratory and at the Museum of Comparative Zoology of Harvard University. For financial support, thanks are due to the Systematics-Ecology Program of the Marine Biological Laboratory, and to the Ford Foundation. I particularly wish to thank the Director of the Program, Melbourne R. Carriker, for his sympathetic encouragement of this work.

Contents

Figures

Introduction: The Problem
and the Sources

In 1859 there began what ultimately may prove to be the greatest revolution in the history of thought. *The Origin of Species*, published in November of that year, effected an immediate and cataclysmic shift in outlook, casting into doubt ideas that had seemed basic to man's conception of the entire universe. The idea of evolution, the explanation of its mechanism, and the book which had such profound effect give to Charles Darwin a unique position in the history of ideas. A thinker of such importance has naturally been the object of much discussion, both learned and popular. Materials for the study of Darwin's life are both copious and accessible, hence it is easy to reconstruct the development of his thought. We have his voluminous scientific writings, an autobiography, a remarkably extensive correspondence, marginalia, manuscripts, and the observations of numerous colleagues, friends, and relatives. Perhaps no scientist's life is better documented.

So much examination, upon such solid bases, would appear to leave little room for further investigation. But somehow the record seems incomplete. There is no agreement among scholars on such basic issues as how and when Darwin discovered evolution or natural selection. Darwin's relationships to his predecessors is subject to a mass of contradictory opinion. We have only the most rudimentary knowledge of the interconnections between Darwin's writings on evolution and those on geology, taxonomy, botany, or psychology. Nor is it really understood what qualities of intellect or accidents of circumstance are responsible for the fact that Darwin, rather than others, brought about the revolution which bears his name. Perhaps this is the most challenging question of all: recognizing that Darwin was a scientist of the first rank, we then ask: what made him so?

One might seek an explanation for Darwin's success in his family background and circumstances of life. He belonged to a family of wealth, but one of liberal outlook and scientific interests. Charles's grandfather, Erasmus Darwin (1731–1802), a successful physician as well as a philosophical poet who concerned himself with scientific affairs, was well known for his evolutionary speculations. Robert Waring Darwin (1766–1848), also a successful physician, was able to leave his son, Charles Robert Darwin (1809–1882), sufficient income to pursue a scientific career without having to work for a living. After his marriage to a wealthy cousin, Darwin became an invalid and almost a recluse, owing to a disease of unknown nature and origin. It is hard to determine whether this illness favored or hindered his scientific work; it limited his time for active labor to about four hours a day, but he seemed to compensate by much thinking and by methodical working habits.

One seeks in vain for any element in Darwin's formal education that might account for his ultimate success as a scholar. He did poorly in the traditional curriculum, dropped out of medical school, and studied without distinction for the ministry. While at Edinburgh and Cambridge, however, he did come into contact with a number of scientists, and was able to indulge his interests in natural history during his spare time. It may be that a distaste for traditional studies predisposed him to think along novel lines. Darwin's characteristic independence of thinking may have resulted largely from the fact that in science he was for the most part self-taught. His real education as a scientist was provided by his experience during five years (1831–1836) as "naturalist" on H.M.S. *Beagle*. Thanks to the assistance of J. S. Henslow, professor of botany at Cambridge, Darwin was able to participate in an expedition which had been organized primarily for the purposes of surveying the coasts of South America and making a series of chronometrical observations. He had for instruction only his books, his instruments, and his own ingenuity. Nonetheless he was most successful, for he returned from the voyage with a wealth of significant geological theory and observation, as well as a foundation for his subsequent biological research.

But these explanations provide only a partial answer to the questions surrounding Darwin's career. His circumstances alone are inadequate to explain his success, and it is not obvious just what his self-education provided. Many have felt that Darwin's intellect, if above average, was not so greatly superior that it could provide the rest of the answer. They have, therefore, been inclined to denigrate his accomplishment and to contend that his innovation was by no means so remarkable as we might think. They argue, for example, that Darwin was lucky—that he stumbled on his discoveries, or that the times were "ripe." But luck does not account for much of Darwin's best work, such as his definitive monograph on the systematics of barnacles. One might argue, and some have, that Darwin was a plagiarist, who took over the ideas of his predecessors without giving due credit.[1] Although such a charge has been made to appear plausible, it fails when the entire body of Darwin's works is taken into account. He could not possibly have stolen his novel and brilliantly reasoned explanation for the origin of coral reefs. Indeed, certain of Darwin's discoveries, namely the curious sexual relations of barnacles and of certain plants, were so original and unexpected that scientists of his time even attributed them to his imagination. It has been maintained that Darwin was a mere naturalist—a remarkably good observer who stumbled on many important facts. Or perhaps he was a man of average intellect who worked hard. But such views cannot be upheld with any rigor in the face of Darwin's excellent experimental work, especially that on plant physiology. Nor can one explain away as mere observation the quantity of theory that pervades his works.

It seems, indeed, a general rule that a consideration of Darwin's writings as a whole casts widely accepted opinions into doubt. Much of his work seems downright anomalous. Why, for instance, did Darwin drop his investigation on the origin of species and spend eight years doing a taxonomic revision of the barnacles? It seems odd that, in later years, he expended so much of his limited energy performing curious experiments on plants and earthworms. Was it really, as it is occasionally said, merely to rest from his more strenuous activities? Before we can be satisfied with any explanation for Darwin's success, we must ask ourselves whether or not the entire sweep of his accomplishment is coherently and consistently illuminated.

Then there is the idea that Darwin was a master of the scientific method—that he was a theoretician of the first rank, a thinker of both originality and rigor, and a philosopher of no mean competence. Opponents of Darwin's theories are inclined to reject this thesis, but there is much positive evidence to support it. Mayr has pointed out how modern Darwin was in his method—that he habitually worked with models and hypothetico-deductive techniques.[2] De Beer in the same vein draws attention to such remarks of Darwin's as this: "The line of argument often pursued throughout my theory is to establish a point as a probability by induction and to apply it as hypotheses to other points and see whether it will solve them." [3] Unless one understands this—that Darwin applied, rigorously and consistently, the modern, hypothetico-deductive scientific method—his accomplishments cannot be appreciated. His entire scientific accomplishment must be attributed not to the collection of facts, but to the development of theory. Both the superficially attractive "true Baconian method" and induction by simple enumeration fail to account for what scientists actually do. Only in the minds of journalists and metaphysicians do significant advances in science result from a determination to be "objective" and to gather facts without any preconception as to what the facts may mean. Indeed, the very act of perception necessarily imposes some kind of order on the course of investigation. That Darwin realized the great importance of hypothesis in his work can be documented by his numerous remarks on that subject.[4] In a letter to a colleague, he explicitly compares his hypothesis of natural selection to the undulatory theory of light with its ether, and to the attractive power in Newton's theory of gravitation.[5] Darwin conceived of his theory as an explanatory hypothesis, as may be seen from another letter, in which he gives the following reasons for supporting it: "(1) On its being a *vera causa*, from the struggle for existence; and the certain geological fact that species do somehow change. (2) From the analogy of change under domestication by man's selection. (3) And chiefly from this view connecting under an intelligible point of view a host of facts." [6]

Thus, Darwin was quite inclined to speculate and form hypotheses, albeit insisting upon verification through experiential tests. His attitude toward this matter is obscured in some places, and certain of his statements could be interpreted to mean that he was an "inductionist." But the following is unequivocal: "False facts are highly injurious to the progress of science, for they often endure long; but false views, if supported by some evidence, do little harm, for every one takes a salutary pleasure in proving their falseness; and when this is done, one path towards error is closed and the road to truth is often at the same time opened." [7] This view is strikingly modern in tone. In recent years, thanks especially to the work of Popper, the idea has become increasingly popular that science advances for the most part not by amassing evidence in favor of hypotheses, but rather by attempts to refute them.[8]

The facts thus seem to indicate that Darwin's philosophical conception of the scientific method was quite sophisticated. An even closer look at the available data reveals that although philosophy has no conspicuous place in his published writings, Darwin was profoundly concerned with philosophical ideas. In his autobiography, Darwin expresses considerable interest in metaphysics.[9] While studying for the ministry at Cambridge, he read Herschel's *Introduction to the Study of Natural Philosophy*, a work which influenced his thinking more than any other.[10] The remarks in his notebooks on works by Whewell are very penetrating, and this can only mean that Darwin read them with critical understanding.[11] Herschel and Whewell were among the most influential philosophers of science of the times, and Darwin knew both of them personally while he was still a young man.[12] But Darwin's philosophy was akin to that of a pragmatist or a logical positivist. It is therefore not surprising that those who adhere to other schools of philosophy have criticized Darwin as philosophically naïve or incompetent. Likewise, empiricists—in the most pejorative sense of the word—have tended to assume that Darwin was as opposed to speculation as they themselves were. Both points of view lead to grievous misconceptions of Darwin and his work; and owing to such attitudes, many have overlooked the extent to which Darwin's writings abound in implicit philosophical discussion.

Given the thesis that Darwin's success resulted from the nature of his method, and given some hints as to the kind of methodology Darwin employed, it should be possible to discover, in his works, both his actual manner of investigation and the sequences of events that led up to his discoveries. In the present study, an effort has been made to treat the general flow of ideas throughout Darwin's works and to construct a single theoretical system which will explain the major features of all of his significant investigations. To accomplish this task is no simple matter of recording the obvious. The many controversies in Darwinian studies often resolve into problems of historiographic technique. An understanding of the scientific issues involved, including some consideration of how knowledge has developed since Darwin's times, is essential to any satisfactory analysis of his thought. The extensive literature on Darwin must be evaluated, not only from a historical and a scientific point of view, but from a philosophical one as well. Much that has been written is based upon notions that are not stated explicitly, and to evaluate such literature it is necessary to understand the basic philosophical and theoretical premises upon which it rests. For all these reasons, it is essential that the evidence be considered in great detail and that the justification for each inference be stated in explicit terms. The result is a work that gives far more consideration to methodology than is usual in strictly historical studies. Not only does this approach do justice to the subject, but it also fulfills my desire to restructure our interpretation of evolutionary biology in the light of the current revolution in the historiography of science and to further that revolution itself.

The bulk of the present work comprises an analysis of Darwin's concepts and an effort to reconstruct the sequence of, and rationale behind, his thinking. A means of verifying the ideas considered here has been provided by the orderly nature of the process of investigation. Hypothetical reconstructions of Darwin's various theories and views have been formulated, and conjectures have been made about the possible sequence of reasoning through which he might have gained insights or reached his conclusions. By considering a number of possible explanations, often by evaluating contrary hypotheses, it has been possible to verify such speculations on the basis of their implications. The elaboration of a hypothesis to explain one point has often been seen, upon reflection, to illuminate previously unsolved problems. Thus, when we see that Darwin approached a problem with a particular set of theories in mind, the reasons become clear not only for his discoveries, but for his errors, oversights, and omissions as well. Disputes between Darwin and other workers become more intelligible when we see that the arguments were based upon different theoretical assumptions and hypotheses. Perhaps most revealing of all has been the analysis of some peculiarities of Darwin's thought processes. Many passages are seen in a new light when we realize that in a variety of situations Darwin made errors which can be explained as instances of the same kind of mistake in formal logic. And a coherent picture emerges of his heuristic—of his logic and method of problem-solving—allowing us to see why particular ideas seemed peculiarly important to him. It is significant that one theoretical system after another has basically the same logical structure. And an understanding of the logic of the theories casts a flood of light upon the history of their discovery. Although this kind of technique has not answered all the questions that might be raised, it does give a certain rigor to the investigation. We know when hypotheses have been refuted, and it is obvious when a large body of facts have been consistently explained. Even where no definitive answer has been given, we have some idea of where to look in future investigations. Thus, understanding Darwin becomes for us largely a question of methodology, much as insight into the causes of evolution could not have arisen for Darwin without an appropriate means of thinking about it.

Anyone who attempts to deal with the history of biology must face the fact that much of the literature is untrustworthy. Many of the secondary sources are based upon metaphysical ideas such as vitalism or teleology which give a severely distorted conception of the biological issues. Often they presuppose the truth of scientific theories which were long ago discarded. Or they view history naïvely, as if knowledge has evolved through mere amassing of fact, or as if a theory can be evaluated apart from the context in which it was used. This is especially true of the literature on Darwin and evolution. In reading the standard histories of biology, one gets the impression that Darwin's work has been largely superseded and that it was full of basic errors of reasoning.[13] This unfortunate condition arose in part because, during the period up to 1930, various reactions and counter-hypotheses were raised against Darwin's theory of natural selection. To some extent these contrary views were justified by the difficulties of the scientific problem. The situation changed completely, however, with the publication of R. A. Fisher's *The Genetical Theory of Natural Selection* in 1930. This work gave natural selection so firm a mathematical basis that within a few years a "synthetic theory" had emerged. Natural selection, Darwin's fundamental contribution, is now as well established as any hypothesis in all the natural sciences. Nonetheless, much biological history was written—and it is remarkable that some continues to be written—from the older point of view. The false premises upon which many writers have based their interpretations affect virtually everything they say. Many such errors have been perpetuated by those who have not gone back to the original works. To find out even the most straightforward facts, therefore, it is dangerous to trust secondary sources.

The Darwin literature is as vast in quantity as it is various in quality. Fortunately for those who wish to review the subject, the significant works form but a small proportion. The standard reference for Darwin's life is *The Life and Letters of Charles Darwin, Including an Autobiographical Chapter*, edited by his son Francis Darwin, who was a distinguished plant physiologist and therefore quite capable of dealing with the scientific aspects of the work.[14] Partly because Darwin lived such an isolated existence, he corresponded extensively; his published letters, many of which are exceedingly useful to the historian, fill several volumes.[15] The many other biographical works on Darwin all suffer to some extent through having been written from an excessively limited point of view. Historians have rarely possessed sufficient background in science to deal with the more recondite biological or geological issues, and scientists have failed to utilize sophisticated historiographic techniques. Among books written by historians, the most nearly satisfactory is Irvine's *Apes, Angels, and Victorians*.[16] A number of works provide useful, and on the whole reliable, information on the historical background of evolutionary thought.[17] The works of biologists, of which perhaps the best is that of De Beer, are on the whole more nearly adequate in their treatment of scientific issues.[18] The reader would do well to beware of a number of recent books which attempt to discredit Darwin's intellect or character.[19] Although specialists in the history of science have had no trouble with this kind of literature, it does pose a problem for the general reader.[20] One needs to compare the citations of sources with the sources themselves before it becomes obvious what such authors have done.[21]

Darwin's research and publications may be separated, for convenience, into six major themes: natural history; the geological; the zoological; the strictly evolutionary; the botanical; the psychological. This division serves to delimit major interests, groups of publications, and sequences of work. Each theme corresponds in part to a period in Darwin's life, insofar as he tended to shift his emphasis from one subject to another. There is a great deal of overlap, however; for example, he retained an interest in geology throughout his scientific career.

The first, or natural history, theme received its greatest emphasis when Darwin was gathering observations and specimens more or less as the opportunity presented itself, without any real attempt to collect evidence for comprehensive theories. Natural history dominated his work during the early stages of his voyage on the *Beagle*. In tracing Darwin's intellectual development from the beginnings of his scientific research in South America, we see that he gradually became more and more interested in theoretical issues, especially in geology. Throughout the voyage, Darwin collected a large number of zoological and botanical specimens and observed the fauna and flora at great length. Yet his biological work had little theoretical content except for biogeographical considerations and an occasional speculation on the nature of species. To the natural history theme we may assign his *Naturalist's Voyage*, first published in 1840, with a second edition appearing in 1845. This work is a popular account of his travels, begun as a journal and running narrative for his relatives. Early in the trip he decided to publish it, evidently taking the writings of Humboldt as a model. It is not strictly a scientific book, and in any attempt to use it as evidence for the development of Darwin's thought, one must proceed with great caution. The book is full of entertaining observations, but contains little explicit theory, even when, as in geology, we know that he was hypothesizing on a vast scale. But it is revealing to compare the version written on board ship, the so-called "Beagle Diary," with the published editions.[22] He did add many points after his return, and in general these relate to significant elements in his evolutionary thinking. The 1840 edition was produced when Darwin had already become convinced that evolution had occurred, but before he had read Malthus on population and had developed his explanatory hypothesis of natural selection. The 1845 edition was modified to include an oblique reference to Malthus, and to add a Lamarckian explanation for the degeneration of eyes in burrowing rodents.[23] Other works of the natural history theme include a few minor papers and the volumes which he edited on the zoology of the voyage, which include Darwin's biogeographical observations.[24]

The geological theme overlaps most with that of natural history. Works are here assigned to the geological group only when they relate to one of Darwin's theoretical systems of geology. These include three books based on the voyage of the *Beagle: The Structure and Distribution of Coral Reefs* (1842); *Geological Observations on the Volcanic Islands Visited During the Voyage of H.M.S. "Beagle"* (1844); and *Geological Observations on South America.*[25] There are also a large number of papers, including several on glacial phenomena, some of which treat work done after the voyage.

One may also abstract a zoological theme, which has some affinity to the next, or evolutionary, theme as well as to that of natural history. Except for some short papers, the works placed here are restricted to the various sections of his monograph on barnacles, published in 1851 and 1854.[26]   Darwin worked on barnacles from 1846 to 1854, although he spent much of his time reading and thinking about evolutionary problems.

The strictly evolutionary theme began to dominate Darwin's research and writings from the time he had completed his study of the barnacles. His great work of this period, *On the Origin of Species by Means of Natural Selection, or the Preservation of Favoured Races in the Struggle for Life,* was first published in 1859, and went through a total of six editions (1859, 1860, 1861, 1866, 1869, and 1871).[27] It was written to take the place of much longer work which Darwin was preparing on the same subject when he learned that his basic hypothesis had been discovered independently by Alfred Russel Wallace (1823–1913). It appeared in November, 1859, having been composed in some haste after the reading of a brief joint communication by Darwin and Wallace to the Linnean Society on July 1, 1858.[28] Notes and manuscripts written prior to 1858 have been published, giving valuable insights into the development of Darwin's thought. Repeated reference will here be made to his notebooks on transmutation of species, which were begun in July, 1837, and extend through the period when he formulated his hypothesis of natural selection.[29] Further developments may be traced through two manuscripts, here referred to as the "Sketch" of 1842 and the "Essay" of 1844, which were both published posthumously.[30] Darwin's second major evolutionary work, *The Variation of Animals and Plants Under Domestication*, was first published in 1862, with a second edition in 1874.[31] His third, *The Descent of Man, and Selection in Relation to Sex*, appeared in 1871, with a second edition in 1874.

After Darwin had finished with the barnacles, he devoted a considerable portion of his time to work on plants. These studies grew out of his evolutionary work and have evolutionary implications, but they display an increasing concern for strictly physiological problems. The previous literature on Darwin has often used these works to illustrate his skill as an experimentalist. The present analysis seeks to demonstrate how they reflect his insight into the broader significance of his evolutionary theories. Besides a large number of papers, we may enumerate six major botanical works: *On the Various Contrivances by Which British and Foreign Orchids Are Fertilised by Insects* (1862); *Insectivorous Plants* (1875); *The Movements and Habits of Climbing Plants* (1875); *The Effects of Cross and Self Fertilisation in the Vegetable Kingdom* (1876); *The Different Forms of Flowers on Plants of the Same Species* (1877); and *The Power of Movement in Plants* (1880).[32]

A final stage or theme is the psychological. Like his botany, Darwin's psychology developed out of his evolutionary thinking. His work was, among other things, a means of studying the evolution of behavior. Arguments of a behavioral nature are of great importance in *The Descent of Man*. A *Posthumous Essay on Instinct*, abstracted from the work which *The Origin of Species* was written to replace, shows that Darwin had long been interested in the evolution of behavior.[33] Even his botany was not without its "psychological" component, in that his studies on plant movements and carnivorous plants are largely concerned with responses to stimuli. Darwin's major psychological work, originally intended as a chapter in *The Descent of Man*, was published in 1872 as *The Expression of the Emotions in Man and*

Animals.[34] There is also his "curious little book" on earthworms, published in 1881 (the year before Darwin died), entitled *The Formation of Vegetable Mould Through the Action of Worms, with Observations on Their Habits.*[35] This work is largely concerned with the "mental powers" of worms, but treats also of their ecological and geological importance.

This division into themes, although not strictly corresponding to a temporal succession, does serve to place Darwin's investigations in a meaningful developmental sequence, one which will be followed in more or less the same order in the following chapters. From the titles of Darwin's many books and papers one could get the impression that he dealt with a medley of more or less independent subjects. The modern university curriculum, so subservient to the needs of specialists, tends to deemphasize the interrelationships between such diverse fields as geology, evolutionary theory, plant physiology, and psychology. Nonetheless, it will here be argued that Darwin's evolutionary biology grew out of his geology, and that essentially all of his work in other fields of biology was the direct consequence of his insights into the nature of evolutionary processes. In effect, the entire corpus of Darwinian writings constitutes a unitary system of interconnected ideas. It strives, with astonishing success, to encompass all organic phenomena within the structure of one comprehensive theory. Only when this greater synthesis is understood, can one really appreciate Darwin's aspiration, insight, and accomplishment.

1. Geology

In answer to a questionnaire sent by Francis Galton (1822–1911), his cousin and the author of *Hereditary Genius*, Darwin wrote, "I consider all that I have learnt of any value has been self-taught." [1] He was, however, only expressing an emotional reaction toward his formal education, for he obviously learned a great deal from discussions with his associates and from such experiences as a long field trip in 1831 with the geologist Adam Sedgwick. Darwin's education as a scientist was mainly provided by his research during the voyage of the *Beagle*, an experience which he considered by far the most important event in his life. [2] He took with him the first volume of Charles Lyell's (1789–1875) *Principles of Geology*, at the suggestion of Henslow, who nonetheless warned Darwin not to believe it. A second volume reached him en route in Montevideo. Thus, although he had no teacher, he did have a celebrated textbook. Provided with talent and a determination to learn, he had only to go into the field, observe, think, and observe again. There is, indeed, no other way to learn science. One must speculate, hypothesize, and make predictions. Most important, one must undergo the painful process of discovering the reasons for inevitable mistakes and failures. For only by knowing that a prediction has proved erroneous, and then reexamining basic premises and logic, is it possible to master the techniques of putting meaningful questions to nature. A teacher or a textbook can be of great assistance, but except in the acquisition of facts (and at this stage method is far more important), learning cannot be a passive endeavor. It was this active process of testing theoretical notions against concrete experience that made Darwin a scientist. No teacher was necessary, and it was an education of the first rank.

LYELL AND UNIFORMITARIANISM

Lyell's ideas were fresh and controversial when Darwin first read of them.[3] Championing the uniformitarian approach of Hutton and Playfair, Lyell opposed, in systematic fashion, the then popular catastrophic geology. He insisted on interpreting the past solely in terms of processes known to occur at present and rejected vast and sudden transformations such as universal deluges. The importance of this step to the methodology of a then young science can scarcely be underrated. Uniformitarianism, by vastly increasing the number of logically possible observations which could refute a geological hypothesis, imposed a new level of rigor.[4] When it is permissible to invoke some extraordinary cause, whenever theories fail to explain the facts, no one will question the basic premises of the theories themselves. And without that repeated criticism which purges the system of its erroneous elements, stagnation all but inevitably results. Therefore it greatly aided the development of geological theory to begin the search for explanations solely with those processes which surely do occur. It might be objected that what is gained in rigor will be lost in rigidity—that real catastrophes might be overlooked. But such difficulties can scarcely be said to have impeded the growth of geology. What mattered was not the exclusion of events and processes alien to our present experience, but consistent and rigorous reasoning. Lyell freely admitted certain types of rapid change, such as earthquakes and local floods. If there have been catastrophes or unfamiliar kinds of changes, these could be discovered after the more familiar phenomena had been understood. The principle of uniformitarianism is thus an instrument of investigation, not a scientific theory. Such charges as the notion that it is based upon a metaphysical assumption that nature makes no leaps must be rejected as absurd.[5] Uniformitarianism is not an empirical proposition at all, and is therefore neither true nor false. We might as well object to discounting unrepeatable experiments on the grounds that we have no reason for excluding the possibility of miracles.

Darwin owed a very real debt to Lyell, both in factual material and in the inspiration for many of his ideas. Indeed, Darwin went so far as to assert "I never forget that almost everything which I

have done in science I owe to the study of his great works." [6]
There is abundant evidence that many of Darwin's theories, both
geological and biological, arose in his mind because he read
Lyell.[7] Darwin was an active supporter of uniformitarianism, and
his publications were of great importance in showing the defects
of catastrophic geology.[8] But the similarities between the two
workers should not be exaggerated. Darwin did not simply expand and illustrate Lyell. Both were original thinkers who made
basic contributions of a quite different nature, and whose styles
of investigation differ markedly. Lyell's contribution was in
formulating a coherent, rigorous, and straightforward methodology and in showing how effective it could be in making a wide
range of phenomena intelligible. His particular explanations are
generally fairly simple. They compel assent, yet they do not
require great power of abstract reasoning to be understood, however ingenious they may be. Indeed, the very simplicity of the
arguments makes it difficult to oppose his views; one cannot
easily rationalize them away. Lyell was also a man of great erudition, able to show effectively the comprehensive scope of his
theories and the fruitfulness of his approach. Darwin, on the
other hand, was another kind of thinker. He was a speculator,
who tended to formulate intricate and subtle hypothetical systems. He was a great methodologist—indeed, one of the best—
but he was not concerned with method as such. Rather, he sought
to employ his methodology in the investigation of specific problems. His approach was basically intellectual, depending on abstract thinking to an extreme degree. His interest in facts lay not
in the illustration of principles but in the testing of hypotheses. In
the degree to which he emphasized the hypothetico-deductive
approach, he differed markedly from Lyell. Darwin's reasoning
tended to be obscure, and his ideas have been exceedingly controversial, almost as a rule.

GEOLOGY AS A MODEL HISTORICAL SCIENCE

In the present analysis, we shall use Darwin's geological
thinking as a basis of comparison for the rest of his work. In the
logic of verification, there are so many parallels between geology and such other historical sciences as biogeography and comparative anatomy that to understand one such discipline puts one

on more or less familiar ground when dealing with another. Geological methodology is also reasonably straightforward, and it was fortunate for Darwin that he began with geological work, for experience in the necessary types of reasoning would seem to have aided his work in other historical sciences. The progressive development of Darwin's thinking—from geology to biogeography, to evolution, and to evolutionary anatomy—becomes readily intelligible when it is seen how similar were his thought processes in all these fields, and how, in several senses, one problem led to another. And his ability to transfer methodologies and theoretical points of view across disciplinary boundaries is one of the main reasons for his success.

One of the most straightforward, yet fundamental, geological methods, a technique used to great effect by Darwin, is the study of sequences. When rocks are examined in many areas, it is frequently observed that they are laid down in beds, or strata. The individual strata may have different intrinsic properties, such as their content in fossils or minerals. If isolated sets of strata are compared with others in different areas, it may be found that the sequences of intrinsic properties are identical, as when sandstone always lies on top of limestone, and both occur above a layer of lava. Such a comparison may be said to be based on formal properties, or more precisely, on the sequential relations of the intrinsic properties. The simplest explanation for this similarity in formal properties would be that the layers in the various areas were laid down by the same processes. Notions of sequential relations may be formulated in symbols, and to do so will help to illustrate the kind of thinking involved and to elucidate some complications. Suppose we had sets of strata in three different areas. Let the letters A, B, C, D, E, and F represent beds of rock with such different intrinsic properties as characteristic mineral content, and let the vertical sequence be represented by the position on the page, with the uppermost stratum at the top:

	Area I: A	Area II: A	Area III: F
	B	B	E
	C	C	D
	D	E	C
	E	F	B
	F		A

If we compare Area I with Area II, we see that stratum D is missing in the latter. The absence of D could be explained in a number of ways; for example: (1) D was not deposited in Area II, or (2) D was eroded in Area II before C was deposited. The available data do not tell us which, if either, is the correct explanation. Now compare Area I with Area III; they are in the opposite order. A reasonable conjecture would be that one group of rocks had been turned upside-down; but here again, the data given do not allow us to infer which group was overturned.

Considering this same example in more detail, we may note that a mere correlation, without any attempt to explain the causes of past geological events, could serve as a basis for erecting a system of classification for the various strata. By a simple inductive process, involving little more than enumeration of similarities, it is possible to obtain a system of comparisons in which like is grouped with like. But such a nonhistorical classification is of little interest or utility, and fortunately there are ways of inferring the actual historical processes. We can, for example, formulate a hypothetico-deductive system which will tell us if a geological stratum has been overturned. That is to say, we hypothesize that a given bed was overturned, predict on the basis of some law of nature that the rocks will have some property because the bed was overturned, and then observe the rocks for confirmation. In such reasoning we might assume, for example, that some of the beds were deposited on a mud flat. We know that cracks form in the surface of a layer of mud when the water evaporates. If another layer is deposited, the cracks may be filled, later becoming preserved when the mud is transformed into rock. The orientation of the cracks tells us which was the upper surface when the beds were deposited. On the basis of observations such as this, the geologist predicts what kinds of features should be expected if his hypothesis is true; and if his predictions are contradicted by experience, he knows that some error occurs in his premises. Although it is not essential to do so, it is a sound rule of methodology to decide what to expect before the observation is made; otherwise the result may be nothing but rationalizations or *ad hoc* hypotheses. And when we get down to fundamentals, it is clear that the process of verification in sciences generally, whether historical or not, is basically the same. To be sure, there are those who have argued, especially with respect to comparative anatomy, that mere correlation, or the comparison of formal relationships, is the basis of all historical research. With respect to geology, the error of this notion is abundantly obvious; but where, as in linguistics, the laws are not so well understood, a certain superficial plausibility has been maintained.[9] Purely formal properties may be used as a basis of linguistic comparison: for instance, the similarities between *cat, Katze, chat,* and *gato* suggest an historical nexus. Yet in linguistics also, laws of nature—including in this case psychological ones—may enter into our theoretical systems and affect our inferences. For example, words resist a change in pronunciation which makes them homonyms of a "taboo" word such as an obscenity, or else people cease to use them when such a change has taken place—as with a synonym for "donkey" in English. It is a tribute to Darwin's grasp of principles that he understood the logic of this approach very well. In a letter to Lyell he says: "Your metaphor of the pebbles of pre-existing languages reminds me that I heard Sir J. Herschel at the Cape say how he wished some one would treat language as you had Geology, and study the existing causes of change, and apply the deduction to old languages." [10] This declaration of confidence in the power of the modern hypothetico-deductive scientific method in revealing the past contrasts sharply with the view of "inductionists," who believe that the facts must be accepted "as they are." And Darwin's success in a variety of historical fields shows that his expectations were indeed well founded.

DARWIN'S GEOLOGICAL RESEARCH IN SOUTH AMERICA

Many of Darwin's efforts in the early part of his voyage on the *Beagle* were devoted to the analysis of stratigraphical relationships in South America. This research is of great interest, in that it would seem to have provided a clue which led Darwin to develop a new theory explaining the origin of coral reefs. His *Geological Observations on South America* is primarily concerned with an attempt to demonstrate that there has been a widespread, extensive, and recent elevation of the entire southern part of that continent. It was of fundamental importance to Darwin's investigation that he establish the sequence of past geological events, and to do this he had to identify strata and other materials over a wide area. An example of the type of reasoning involved, which employed methods that had been worked out by his predecessors, may be seen in Darwin's work on igneous intrusions. Often he notes how volcanic rock cuts across a number of strata (fig. 1).[11] This usually implies that the volcanic activity occurred after the strata were deposited. An alternative hypothesis might be raised that preexisting volcanic rock was surrounded by water in which deposition of sediments took place. However, Darwin's descriptions of igneous intrusions include remarks on the effect of heat on the adjacent rock (metamorphism), which clearly excludes such an alternative. Given sufficient evidence of this type, the sequences of events could be placed in proper order, so that the various insights ultimately fit together like the pieces of a jigsaw puzzle.

West

East

Flagstaff Hill
2,272 feet high

The Barn
2,015 feet high

FIG. 1 Strata at St. Helena cut across by volcanic dikes (shaded transversely). The dikes must have been formed after the strata. From *Volcanic Islands*.

In attempting to corroborate his hypothesis that South America has undergone widespread elevation, Darwin sought to support his stratigraphic and other observations by erecting a system incorporating several independent lines of evidence. For instance, he gives diagrams showing steplike terraces, or benches, which he found to occur in a series extending far inland over a wide area in South America (fig. 2).[12] He interprets these features as the result of marine erosion during interruptions in the process of elevation.[13] By identifying and tracing the terraces, the sequence of processes involved in the elevation could be determined for the separate areas. Darwin also studied the fossils, in an attempt to gain both another indication of sequences and a sort of crude dating. Time and again, he refers to the fact that the shells on a level far above that of the sea were found to be taxonomically indistinguishable from contemporary species. In view of the fact that there is a gradual increase in the percentage of modern species among progressively more recent deposits, Darwin was able to estimate the relative rates of elevation in different areas. In his progressively higher collections of shells, he observed a gradual loss of color, which once more indicated sequence, because the shell pigments ought to decompose at nearly the same rates. A major earthquake occurred while he was in Chile, and by observing its effects and referring to records of similar events, he was able to show that considerable uplift had recently taken place.[14] The study of earthquakes in turn fitted into generalizations about the nature of elevation and subsidence. The particular kind of evidence which Darwin used in this work is especially significant, for it illustrates one of the most

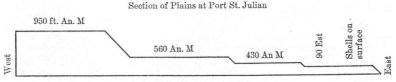

FIG. 2 Terraces: evidence of gradual, interrupted uplift, and a basis of comparison over a wide geographical area. From *South America*.

useful principles in the historical sciences: independent evidence. Terraces, shell taxonomy, pigments, and direct observations on the effects of earthquakes are selected, not merely because they add to the quantity of evidence, but because they give it a different quality. The hypothesis of elevation is consistent with only a very restricted set of conditions predicted in the different phenomena under study. Each line of evidence has a very definite field of application, and the hypothesis is supported by the joint implication of the various arguments.[15] Although any of the separate kinds of argument might be explained away in terms of some *ad hoc* rationalization (such as asserting that the shells had been carried up by Indians), it is far more difficult to make one such rationalization explain away several different kinds of evidence at once. To add a new line of evidence complicates the pattern of relationships which can be derived from the theory being tested. Therefore, invoking a greater variety of arguments constitutes a far more serious attempt to refute the theory than merely increasing the quantity of data; and only an effort at refutation constitutes a real test. Independent evidence, again, is fundamental to all historical sciences. When an historian, in the traditional sense of the word, weights his sources for maximal independence, his action in basically the same logical operation used by a geologist is basing his stratigraphic correlations on several taxonomic groups of fossils, or by a comparative anatomist when he prefers a classification system founded on several organ systems. In terms of methodology, the historical sciences are all basically alike, and the separation of those which deal with man from those which do not is merely an artifact of human vanity.

In considering further some peculiarities of historical sciences, we note that there was essentially no quantification in the geological work described so far. Darwin did measure distances and heights, but these had little to do with his arguments as such. All he needed to demonstrate widespread elevation was relative sequences. Nor did he need to date the particular events which he placed in temporal order. A series of dated events would have been more informative, but it would not have made his inferences more correct. This point may seem obvious, but it is sobering to realize that some persons have criticized comparative anatomy on just such grounds: it does not give numerical dates. And there are those, even some physicists, who maintain that scientific work must employ mathematics. But mathematics can reasonably be treated as a branch of logic, and to view any form of logic as something more mysterious or valid than what is called common sense is without foundation and smacks of the superstition of numerology. Scientific inferences should be accepted because the premises are true and because the conclusions follow logically. The truth does not derive from the jargon in which it is expressed.[16]

CORAL REEFS: ORIGIN OF THE HYPOTHESIS

The discovery of widespread elevation in South America had a profound effect on Darwin's conception of geological processes in general. He realized that, although some areas had risen, others must have subsided, and that both rise and fall should have widespread consequences. He developed his ideas, and by the time he had returned from the voyage of the *Beagle*, Darwin was able to present Lyell with the first of his great syntheses, the theory of coral reefs. Elegant, yet simple, and wholly original, it is perhaps the easiest of Darwin's theoretical systems to understand. Large areas of the Pacific Ocean are strewn with small islands called atolls. These islands are composed entirely of coral and other calcareous deposits, and often have a ringlike configuration, with a central body of water (lagoon). The origin of such islands had long been a mystery. Lyell had explained the ringlike arrangement as a result of coral having grown upward on the rims of submerged volcanic craters. Darwin's theory was far more comprehensive and satisfactory, and won Lyell's immediate approval. Darwin divided coral formations into three types: fringing reefs, barrier reefs, and atolls. Fringing reefs occur adjacent to a body of land. Darwin held that they form when coral grows outward from the land, in the direction of the water. Barrier reefs are a kind of submerged coral wall, at some distance from land and separated from the shore by a lagoon channel. They would arise if the land were to sink and the coral on the outer edge of a fringing reef were to grow upward. If, say, a volcanic island were partially to sink, the coral would grow upward at the outer rim of any fringing reef that surrounded it and would leave a lagoon channel separating a barrier reef from the shore. Should sinking continue, the island itself would ultimately disappear from view. Yet if the coral continued to grow upward on the barrier reef, the circle of coral would be maintained, and the result would be an atoll (fig. 3).[17]

The antithesis between elevation and subsidence forms an obvi-

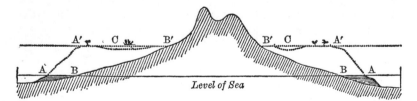

FIG. 3 Stages in formation of an atoll. AB, fringing reef; A'B', later stage, when island has sunk, with barrier reef (A') and lagoon channel (C); A"A", still later stage, with island totally submerged, outer edges of reef (A") surrounding a lagoon (C). From *Coral Reefs.*

ous link between Darwin's work on South American geology and his theory of coral reefs. There can be no serious doubt that this conceptual relationship corresponds to a genetic one, for in his autobiography Darwin has provided us with a clear-cut account:

> No other work of mine was begun in so deductive a spirit as this, for the whole theory was thought out on the west coast of South America, before I had seen a true coral reef. I had therefore only to verify and extend my views by a careful examination of living reefs. But it should be observed that I had during the two previous years been incessantly attending to the effects on the shores of South America of the intermittent elevation of the land, together with denudation and the deposition of sediment. This necessarily led me to reflect much on the effects of subsidence, and it was easy to replace in imagination the continued deposition of sediment by the upward growth of corals. To do this was to form my theory of the formation of barrier reefs and atolls.[18]

With respect to this particular theory, it is evident that the source of Darwin's original idea was a strictly rational development of theoretical concepts. We will see in future chapters that similar processes are sufficient to explain all of Darwin's major contributions, a point which relates to many important philosophical and psychological questions. With respect to *philosophy*, we need to consider the various problems surrounding the relative significance of inductive and deductive reasoning in science. These problems are greatly complicated by the fact that scientific research involves a variety of steps, each of which uses a different kind of reasoning. The problem of induction—of how one justifies theoretical generalizations from individual events—is a perennial issue in the philosophy of science. It is not to be confused with the *psychological* problems of the origin of new ideas, problems which are not recognized as a part of logic. For broader studies on the nature of scientific thought, however, the pragmatic, heuristic, and psychological problems are of the utmost interest. We have for one thing to deal with the notions that creativity involves thinking of a strictly nonrational order or that there is something verging on the supernatural in what has been called "inspiration." Such views are often mere affirmations of ignorance or hope, analogous to the view that science cannot explain "mind" or "life." Beyond the value of casting unscientific attitudes into disrepute, there are many unsolved problems in the mechanics of intellectual innovation, and one would expect a thinker of Darwin's originality to be most worthy of investigation. Insofar as reliable evidence is available, the facts would seem to indicate that Darwin's ideas were generated by an orderly, rational process. More general considerations will be necessary before it can be established how well this compares with other scientists' work, or whether the proposed mechanisms are the only source of new ideas.

The historical or psychological origin of Darwin's ideas through reasoning from preexisting generalizations does not contradict the fact that, in the context of verification, he did not trust the so-called deductive approach. He was most critical, for example, of Herbert Spencer on this point.[19] In his autobiography, he remarks that it was only in his work on coral reefs that he could remember forming a hypothesis without having to modify it greatly, and he gives this as a reason for mistrusting "deduction." [20] But we may note that what he called "deduction" is by no means thus rendered unnecessary; what matters is that the theory must be tested by experience, irrespective of how much "deduction" is involved. Nor should one confuse the reasoning giving rise to a hypothesis with arguments demonstrating that the hypothesis is valid; again, discovery is not verification. The universe appears to be so constructed that we rarely find it possible to elaborate, out of purely theoretical considerations, a system showing as nearly perfect correspondence to experience as did Darwin's coral reef theory. It is usually necessary to reconstruct our systems after a partial, but not necessarily complete, falsification. To students of the scientific method in general, and to those of Darwin in particular, the work on coral reefs is an almost ideal model. It happened that Darwin had all the elements he needed to construct and test his theory, and that no complications forced him to alter his hypothesis. All the elements of this work, a classic example of the scientific method, are isolated, and thus they may be studied separately. It can serve, therefore, as a paradigm for comparative study of his other works, in which the underlying rationale of the various operations is rarely so obvious.

CORAL REEFS: CORROBORATION OF THE
HYPOTHESIS

One way to test a hypothesis is to fit it in with a broader
system of generalizations. This method, typical of Darwin's later
work, was applied to the coral reef problem. Having observed
that in South America elevation and subsidence occur over considerable areas, he reasoned that the products of subsidence
(atolls and barrier reefs) should display a particular kind of distribution pattern. His book on coral reefs includes a map, the
product of an exhaustive survey of the literature, showing the
distribution of different types of coral formation all over the
world. It is clear from the map that atolls and barrier reefs occur
in definite areas, such as off the coast of Australia, and in bands
across the Pacific and Indian oceans. Fringing reefs are likewise
concentrated in particular regions. Within any given area, the
type of reef is quite uniform.

The geographical relationships could in turn be derived from
other theoretical generalizations. In South America Darwin had
observed that active uplift correlates with vulcanism, and that
volcanoes and mountain chains tend to be distributed in bands.
He developed a theory to explain this relationship, involving the
notion that injection of igneous material between strata causes
the overlying rock to rise. Modern geology usually invokes other
explanations for uplift, although it does admit the ability of igneous intrusion to lift rock in such phenomena as the formation of
swells. However, the correlation noted by Darwin appears to be
related to more fundamental causes. The failure of Darwin's explanatory mechanism exemplifies the unreliability of correlations
as evidence for causal relationships, really only a more particular
instance of the shortcomings of induction in general. But just as
Newton's physics, though replaced by Einstein's, retains some
utility, Darwin's views on uplift were not completely wrong; the
correlation stands, and the mechanism has some validity. In his
work on volcanic islands, Darwin observes that the linear orientation of islands in the Galapagos Archipelago is analogous to that
which prevails in other Pacific island groups.[21] His map of reef
types also shows that areas of vulcanism occur in narrow strips
over much of the world, and that the areas of marine vulcanism

often coincide with fringing reefs. Where there is little vulcanism, as in part of the West Indies and certain areas of Africa, the presence of fringing reefs could readily be explained as reflecting a rise somewhat distant from the main area of uplift, like that of the eastern coast of South America. Darwin stresses the fact that "There is not one active volcano within several hundred miles of an archipelago, or even a small group of atolls." [22] On the whole, then, it is possible to demonstrate that atolls and barrier reefs occur in areas where one would expect there to have been subsidence; and where there are fringing reefs, there should be signs that the coastline has risen.

Darwin's coral reef theory, at least in broad outline, is now confirmed by a mass of evidence.[23] The modern view differs primarily in recognizing the added influence of changes in sea level in the alternate growth and erosion of coral structures. Drilling on Eniwetok atoll in 1952 demonstrated that shallow water corals have gradually produced, over a period of some millions of years, a layer of limestone several thousand feet thick.[24] In addition, soundings have revealed that in the same area as the atolls there are numerous flat-topped, yet deeply submerged mountains known as guyots. These have resulted when volcanic cones eroded at sea level and sank without sufficient growth of coral to maintain them. That such structures might exist was implicit in Darwin's theory, and indeed his discussion of a "dead" reef, the Great Chagos Bank in the Caribbean, shows an early stage in the formation of a sunken, flat-topped structure.[25] Yet Darwin's theory of coral reefs, like many of his ideas, was long controversial. A number of writers argued against it on grounds which we now see were irrelevant from the beginning. As a general rule, the distributional arguments were overlooked by Darwin's critics, in spite of the fact that these are the basic means of justification.[26] Murray, for instance, completely missed the point in his argument that perhaps coral reefs could be built up without subsidence, as the factual issue was resolved by the argument from distribution.[27] And Darwin's demonstration that the sides of atolls are exceedingly abrupt provides a fact totally inconsistent with the notion that atolls could have been compacted out of soft sediments. In a letter to Alexander Agassiz, Darwin explains the entire problem in temperate language, but Agassiz evidently failed to understand the argument.[28] It has been observed that most criticisms of the theory came not from geologists but from zoologists.[29] Like many controversies in science, this one had nothing to do with the facts; it resulted from basic differences in philosophical outlook which were never really brought out into open discussion.

THE CORAL REEF THEORY AS
SCIENTIFIC EXPLANATION

Darwin's coral reef hypothesis differs somewhat from the usual type of theory in chemistry or physics. Laws of nature and propositions about particular historical events occur admixed in a manner unfamiliar to the less historical sciences. Yet it is a simple, rigorous, and well-tested theory, as deserving of being called "scientific" as any. As it has a certain analogy with other Darwinian contributions, it may help to use it as a standard of comparison for treating various philosophical ideas. Attention to such matters seems justified in view of the fact that what has been called the "philosophy of science" has all but completely overlooked such fields as biology and geology. Let us begin with the question of how a theory explains. It is obvious that Darwin's coral reef theory constitutes an explanation, but it is far from clear precisely what we mean when we say that a theory is explanatory. Philosophers have had a great deal to say on this topic, but their assertions are not very enlightening. One of the more popular concepts is the so-called "covering-law model" or "predictionist thesis," largely supported by Hempel.[30] This view attempts to relate all explanation to deducibility from supposedly invariant laws of nature and, in its more extreme form, to prediction. Such a notion has some merit, in that there is usually a way in which we can relate facts to observed regularities in our experience, and we do tend to look upon such relationships as explanatory. But "prediction" must be used in a nontemporal sense when fitting the thesis to past events.[31] "To follow," in a logical sense, here would mean only that the law implies certain events, a relation having no reference to the position of the observer in time. This justifies a logical equivalence for explaining past events, and predicting future ones, in terms of a law. But it is equally clear that the "covering-law model" must be hedged with many qualifications. The problem becomes even more acute when it is seen that for such historical sciences as geology the predictions are anything but precise. It is misleading to say that when an explanation is requested we ask for reference to laws of nature, descriptions of original conditions, and the ability to know what must happen of necessity. Rather, we ask for only the most approximate relationships between the conditions and the laws. A kind of prediction is essential, but only in the sense of giving a range of probabilities. Thus, Darwin's coral reef theory clearly relates and forecasts, but it does not prophesy. It tells us where we should expect atolls or barrier reefs to occur. Yet no ironclad prediction is made as to whether or not a given mountain rising out of the sea will ever become an atoll. The theory must be expressed in conditional terms: *If* coral grows on an island, *if* the island sinks, and so forth, *then* an atoll will ultimately be formed. And where there is a barrier reef, various conditions have been met, including either subsidence, or a rise in sea level, or perhaps something else. In more philosophical terms, the explanation gives possible necessary conditions, but only when an atoll has been formed do we know that the conditions have been sufficient. One might perhaps say that there is a difference between historical sciences like geology and "hard" ones like physics, but this misses the point. *If* I release an apple, and *if* it is not held up by a string or something else, *then* the apple will fall. Neither Darwin's theory of coral reefs nor any law of physics tells us that the conditions must be met. In fact, to demand anything more than an approximation from the covering-law model is gratuitous.

Geological generalizations frequently seem not to predict in the manner usual for physics because the conditions with which we must deal are so intricate. The reasoning involved has more in common with our operations in driving an automobile; we predict, crudely, our future path, but since we cannot account for every contingency, we must repeatedly supply conditions and remake the calculations. In the context of a historical explanation, emphasis is given to the conditions which have been met, and although the invariant relations, or laws of nature, are present, they are looked upon as less significant than the conditions. Attempts have been made to sweep the problem of historical explanation under the rug by means of a verbal maneuver: statements of conditions are called "description" and derivation from laws "explanation." [32] But obviously there cannot be an explanation without something to be explained, and therefore conditions are indispensable in explanation.

But perhaps the search for a logical treatment of explanation is beside the point. The concept of explanation is in a way not logical but psychological. As Dray puts it, explanation has a pragmatic function.[33] A phenomenon is explained to us insofar as we are put in a position to relate the phenomenon to the rest of our experience. An explanation thus puts both raw fact and theory into a context in which they can be used. Darwin's coral reef theory seems to us far more than a mere description, largely because it is so rich in implications—it tells us where to look for further development of our knowledge. Yet this kind of interpretation also has its difficulties. In our concern with the nature of explanation, we must deal with a fuzzy boundary between the logical and the psychological. But it is easy to confuse the concepts of logic with those of psychology. The two disciplines meet in the field of heuristic, or in the pragmatic aspects of problem-solving. Simply because the boundaries seem indistinct, one may expect all sorts of errors. One of these is psychologism, or the misinterpretation of logical ideas by confounding them with psychological analogues. We saw this kind of mistake in the distinction between the discovery of a hypothesis and its justification. In addition, we may note an analogous attempt on the part of logicians and others to misinterpret ideas properly psychological as if they were the same as logical ones. For instance, logical positivism has developed a special usage for the word "meaning," to refer to strictly empirical relationships: a proposition is meaningful only if it can be verified. It is easy to let this concept intrude into the wrong contexts. To insist upon using the word "meaningless" as a blanket term for propositions not verifiable by the methods of science tends to be a propaganda device for degrading the very meaningful (in another sense) realms of aesthetic and religious experience. If we are to understand science, we cannot overlook the scientist's thought processes. It is here that we encounter difficulties such as psychologism, and the pitfalls are to be avoided only by drawing the appropriate distinctions. And for much the same reason, studies in "pure" logic tend to have little relevance to a comprehension of actual processes of scientific investigation carried out by real human beings. The difficulties surrounding explanation cannot be solved merely by redefining the word so as to please logicians, for a strictly logical interpretation may be beside the point. Philosophy becomes a mere exercise in logic-chopping when cut off from experience, but it makes a truly significant contribution as an integral part of learning in general. The language of science is a living tongue, spoken by flesh and blood human beings with real ideas to convey. And just as biology is more than the study of corpses, a full understanding of science requires that we know how ideas are actually used.

REFUTATION AND SCIENTIFIC EXPLANATION

Among the most important criteria of scientific reliability is the ability to withstand critical testing. Popper maintains, and I think rightly, that a theory is not scientific unless it is possible to refute it.[34] There are any number of widely held notions which do not stand up to this test. Many of these are to be found in the excesses of psychoanalysis: when the pleasure principle fails, one may always invoke the death wish. A theory consistent with all logically possible future experience is clearly only a pseudo explanation. Could one assert that Darwin's coral reef theory is of this nature—that there is no possible way to refute it? This question is most significant, for analogous charges have been raised against Darwin's evolutionary theories. It will be somewhat easier to evaluate the issue with respect to coral reefs because the evolutionary theories, while similar in principle, are far more intricate. It might be argued that the coral reef hypothesis cannot be refuted because, although a mass of facts are available which support it, none conflict with it. But this misses the point, for the test of verifiability rests not with known facts, but with logically possible ones. Surely a valid theory is not likely to have many actual pieces of evidence against it. The charge is unfounded, for one has no difficulty in specifying conceivable observations which would refute Darwin's theory. It is logically possible that we have misinterpreted the data—that atolls do in fact occur mixed with active volcanoes, or in areas of active uplift. To Darwin this seemed a practical possiblility as well; therefore he surveyed all the available literature as a critical test of his theory, and had the facts turned out not to be as he had anticipated, he would have been forced to reject his hypothesis.

A more realistic challenge to the theory would be the discovery that certain, hitherto unknown organisms are capable of building up coral deposits in deep water, and the demonstration, through borings, that this has actually happened. The possibility is remote, but it is still with us.

Another form which the charge of untestability might take would be that for any individual reef one may explain away anomalous forms of coral growth. Or it might be argued that one cannot predict the course of development for a given coral formation: if an atoll forms, it was in consequence of the forces hypothesized; if none forms, then the theory is enlarged to admit of no growth at all. Although such assertions might seem plausible at first glance, consider how the theory is actually tested—by the distributional argument. The hypothesis does not imply that reefs must form in certain places, or that they will assume a given structure, but only that there will be a correlation between the type of reef formed and the local conditions of sinking or uplift. The argument is a statistical one. The objection is based upon the fallacy of division: it assumes that every element in the argument must have the same properties as the argument taken as a whole; it overlooks the fact that the test is of the system, not the parts in isolation. One might as well argue that because the deaths of individual human beings cannot be predicted, there is no justification for the mortality table. But why do insurance companies make money? Analogous misunderstandings about the nature of statistical reasoning underlie many attacks on Darwin's theories, and such problems will be a recurrent theme in later chapters.

Darwin's many other contributions to geology include a substantial body of data from his years on the *Beagle*, his novel theories on the formation of volcanic bombs and on cleavage and exfoliation, his papers on glacial phenomena, his demonstration that the fossil record is far less complete than had previously been thought, and his insights into the mechanism of soil formation. As these topics are fairly well treated elsewhere, and as they have little direct bearing on the problems under consideration, they will not be discussed further in this account.[35] Yet one does well to bear in mind the fact that Darwin's contribution to geology was substantial, and that it alone would justify him an important place in the history of science. It is only by comparison to his work on evolution that Darwin's geology seems unimportant.

2. Biogeography and Evolution

It is well known that both Darwin and Wallace came to discover the phenomenon of evolution through the study of biogeography. This fact takes on added significance when it is realized that both men made important contributions to biogeographical theory and methodology.[1] In *The Origin of Species*, the strongest positive argument for evolution is the geographical one; sufficient evidence was available at the time for a detailed evolutionary treatment of both plant and animal distribution, while the now equally convincing paleontological evidence had only begun to accumulate. It is no accident, therefore, that many of the immediate converts to evolution were younger scientists actively engaged in the study of distribution, particularly of plants. This is true for the American botanist Asa Gray (1810–1888), and even more strikingly for Bentham, who changed his mind immediately upon hearing the joint paper of Darwin and Wallace in 1858, in spite of the fact that he had planned to deliver an attack on evolution the very same day.[2] But the precise relationship of Darwin's biogeographic work to the emergence of his evolutionary thinking has long remained obscure.

There are two schools of thought on the origin of Darwin's idea that species are not stable. T. H. Huxley is largely responsible for one view—which has more recently been supported by Smith and others[3]—that Darwin did not become an evolutionist until *after* he had returned from the voyage of the *Beagle*, when he was able to sort out and study his collections as a whole. A considerable amount of evidence can be marshaled in support of this conception. There is a letter to the German botanist Otto Zaccharias, dated 1877, in which Darwin says: "When I was on board the Beagle I believed in the permanence of species, but, as far as I can remember, vague doubts occasionally flitted across my mind. On my return home in the autumn of 1836 I immediately began to prepare my journal for publication, and then saw how many facts indicated the common descent of species. . . ." [4] There are, in addition, Darwin's notebooks, which assert that the evolutionary hypothesis dawned upon him in 1837, several months *after* his return. The following seems conclusive: "In July opened first note book on 'Transmutation of Species'—had been greatly struck from about Month of previous March on character of S. American fossils—and species on Galapagos Archipelago. These facts origin (especially latter) of all my views." [5]

Other arguments are less convincing. Smith criticizes Darwin for his assertion that he gathered facts by "Baconian" induction, when in fact he had already become a convinced evolutionist.[6] Yet Darwin clearly meant that he had no hypothesis about the *mechanism* of evolution, and in this limited sense his investigation was rather "Baconian." [7] Again, it has been argued that Darwin was not an evolutionist until after his return on the grounds that one cannot find traces of evolutionary thinking in the "Beagle Diary." [8] It is true that the cryptically evolutionary passages in the published version were added at a later stage. But the evidence is purely negative. Why should Darwin discuss speculations about the nature of species in his diary? There is very little geology in the diary—only enough to describe the local scenery. The diary is not basically scientific, but rather was designed for entertainment, to recount his adventures to his family, to recall his experiences in later life, and to serve as the basis for a popular work. Such a journal is no place to record vague and technical speculations. Nor is there any reason to think that he would put such ideas down on paper until he began their methodical development.

The contrary view of an earlier origin for the evolutionary hypothesis was first supported by Darwin's son Francis.[9] While not denying that greater certainty was developed after the voyage, the "vague doubts" expressed in the letter to Zaccharias are given more weight. Darwin's rather poor memory (a ready-made instrument for explaining away unpleasant facts) is invoked to justify the notion that he was converted to evolution while still on board the *Beagle*, perhaps subconsciously and only tentatively.

In support of this view, Barlow cites the following passage from Darwin's ornithological notes, where he discusses his impressions of the Galapagos:

> When I see these Islands in sight of each other, & possessed of but a scanty stock of animals, tenanted by these birds, but slightly differing in structure & filling the same place in Nature, I must suspect they are only varieties. The only fact of a similar kind of which I am aware, is the constant asserted difference—between the wolf-like Fox of East & West Falkland Islds. —If there is the slightest foundation for these remarks the zoology of Archipelagoes—will be well worth examining; for such facts would undermine the stability of species.[10]

This is, at least, a real questioning; yet it does not of itself make Darwin a convinced evolutionist. And later statements in the same notes are quite hypothetical in tone when they refer to instability of species. A number of other quotations have been cited as ostensibly supporting the notion that Darwin was an evolutionist during the voyage.[11] Perhaps the most celebrated of these is in the opening lines of *The Origin of Species:*

> When on board H.M.S. "Beagle," as naturalist, I was much struck with certain facts in the distribution of the inhabitants of South America, and in the geological relations of the present to the past inhabitants of that continent. These facts seemed to me to throw some light on the origin of species—that mystery of mysteries, as it has been called by one of our greatest philosophers. On my return home, it occurred to me, in 1837, that something might perhaps be made out on this question by patiently accumulating and reflecting on all sorts of facts which could possibly have any bearing on it.[12]

These statements seem most convincing. But another look—and we must bear in mind the fact that Darwin tended to write in a rather ambiguous style—raises some doubts. He was, during the voyage, "struck" by facts. They "seemed" relevant to the origin of species. But how were they relevant? To what degree? Were they more than suggestive? And when did they "seem" relevant? We cannot with any rigor infer, from the statement itself, that the facts at any particular stage made an evolutionist of him. Nor should we forget that we are here dealing not with a purely historical account, but with a dialectical maneuver. In arguing for evolution it was natural to stress the "inductive" aspects of the discovery—to imply, that is, that his insight arose not from speculation but from the facts. Darwin, like other scientists of his day, gave much lip service to "induction," and such hypocrisy has long obscured the real nature of scientific discovery.

Actually, the proper distinctions allow one to reconcile the apparent contradictions in the various documents. Darwin evidently meant that during the voyage he was struck by certain observations, without at once being aware of their full evolutionary significance. We know from documentation which facts impressed him. Judd enumerates letters home and other evidence which shows quite clearly that Darwin was fully aware how important were certain mammalian remains he had unearthed in the Pampas.[13] He had discovered a number of large fossil mammals which differed conspicuously from the present inhabitants of the country. In association with these, he found the remains of mollusks no different from the contemporaneous forms. The inference would be inescapable that within quite recent times some of the local species had vanished while others had remained unaffected. A clear connection had therefore been made in Darwin's thinking, between the sequence of geological events and changes in the biota. There can be little doubt that he did in fact relate his timing of events by the study of shells to his work on distribution, as he makes casual reference to this relationship in the "Beagle Diary." [14] On the other hand, there is no real evidence that his biogeography was, during the voyage, an evolutionary one. His notes, especially those written during the latter part of the journey, do refer to distribution, centers of creation, invasion, extinction, and, hypothetically, even the possibility of species change.[15] There are remarks upon the distribution of animals in the Falkland Islands or southern South America; especially revealing are statements that particular areas are devoid of certain forms present on the mainland, suggesting that their absence is due to a barrier preventing their invasion.[16] Yet it is all too easy to read an extraneous evolutionary intent into the sources. Comparative anatomy, now permeated with evolutionary thought, was worked out with no assumption that there is an underlying historical nexus. The same is true, albeit to a far lesser extent, of biogeography. One may elaborate a conception of biogeographic realms or provinces without treating the patterns of distribution as the result of historical changes. The older biogeography was largely based upon centers of creation, and there are no real grounds for holding that Darwin organized his thought in terms of any other principle until after he had returned to England. An altogether different perspective on the problem was necessary before this could happen, and the essential elements of an evolutionary point of view were still lacking.

THE INVENTION OF
EVOLUTIONARY BIOGEOGRAPHY

During the *Beagle* expedition, Darwin's biogeographic thinking was traditional. It was organized around the notion of centers of creation, with some admixture of the ecological approach typical of Humboldt. His manner of thinking is made clear from an entry in his "Beagle Diary" of August 5, 1834, in which he describes the "aspect" of the country near Valparaiso, then comments:

> With this sort of vegetation I am surprised to find that insects are far from common; indeed this scarcity holds good to some of the higher orders of animals; there are very few quadrupeds, & birds are not very plentiful. I have already found beds of recent shells, yet retaining their colors at an elevation of 1300 feet; & beneath this level the country is strewed with them. It seems a not very improbable conjecture that the want of animals may be owing to none having been created since this country was raised from the sea.[17]

This passage, which does not occur in the *Naturalist's Voyage*, is important both because it shows that Darwin was still thinking like a creationist rather than an evolutionist, and because it links together his thinking on both geological and biological history. At this stage, distribution was explained as the result of the pattern of localized special creation, and Darwin sought to answer such questions as when and where creation and extinction had occurred. The same outlook was still fundamental to his thinking at the Galapagos Islands, for he says: "I industriously collected all the animals, plants, insects & reptiles from this Island. It will be very interesting to find from further comparison to what district or 'centre of creation' the organized beings of this archipelago are attached." [18] Even later on in Australia, an entry in the "Beagle Diary" gives reflections on the differences between the local animals and those elsewhere, in terms of it appearing that there had been "two distinct Creators," although it seemed curious that the European and Australian ant-lions were generically identical.[19] Here he stared the answer right in the face, but did not see it—a critical element had yet to be supplied. Nonetheless, the failure of special creation to account for many of the observed phenomena evidently predisposed Darwin to seek a better explanation, and there is abundant evidence that evolution had at least occurred to him.

One more link in the conceptual chain had to be provided before the evolutionary interpretation was inescapable. The key insight seems to have arisen because of observations on the invasion of islands by organisms introduced by man, observations which became fairly frequent toward the end of the voyage, especially in New Zealand, Australia, and St. Helena.[20] When it had been observed that introduced animals or plants may spread rapidly in new areas, the way was open to speculation upon what would happen if natural populations were to invade new territory, and to asking what prevented their dispersal. It would be clear from past experience that barriers—mountain chains and oceans, for example—are able to limit the spread of organisms. Yet what would constitute a barrier for one form would be nothing of the sort for another. The pattern of distribution is determined for each form by its type of locomotory mechanism and by the particular barriers which it happens to encounter. And once one sees that remote islands, such as the Galapagos, are populated by forms particularly well adapted to crossing the prevailing barriers, while the local barriers would account for local diversification, an evolutionary interpretation is inescapable. This particular argument is a dominant theme in both the notebooks on transmutation of species and *The Origin of Species*.[21] Once the crucial insight had been gained, but not before, it would be obvious that the peculiarity of Australian mammals resulted from an inability to cross water masses, while the ant-lions were able to colonize Australia by flight.

Whenever it was that the insight came to him, Darwin, in seeing the connections between dispersal mechanism, barriers, and patterns of distribution, had discovered the definitive evidence for evolution. Nobody had ever done this before, and it is the second of Darwin's great syntheses. Even Lyell, who had considered biogeographic problems along analogous lines, failed to see the evolutionary implications.[22] Those who seek to deny Darwin credit as the founder of the theory of evolution, on the grounds that others had previously thought of evolution or argued for it, misconstrue the nature of scientific innovation. Before Darwin, evolution was a mere speculation, one which could be invoked to explain some facts, but which was unsupported by any truly critical tests. With Darwin's biogeographical synthesis, evolution became an integral part of a compelling system of explanatory theory. A new perspective on the data had emerged. The absence of all mammals other than bats on remote islands, or the wealth of flightless birds in New Zealand, could no longer be looked upon as a brute fact.[23] The basic pattern could never again be ignored; either evolution had to be accepted, or the facts had to be explained away.

The fact that the general features of distribution can hardly be interpreted otherwise than as the result of evolutionary processes forms one of the strongest arguments in *The Origin of Species*. The evidence is marshaled with great strategic effect, in two chapters which compare, point by point, the relative merits of evolutionary and creationist interpretations.[24] Likewise, Darwin takes the utmost pains to show the insufficiency of the contrary hypothesis of an ecological explanation for the distribution of taxonomic groups. The reasoning was well developed early in Darwin's investigation; it forms a major portion of his first notebook on transmutation of species (1837–1838), and therefore preceded his discovery of natural selection.

FURTHER ELABORATIONS OF BIOGEOGRAPHY

Darwin evidently realized that developing a comprehensive system of biogeography would show the great utility of the evolutionary theory as a guide to scientific investigation. He devoted much of his time, especially before *The Origin of Species* was published, to a detailed study of the subject. He never got around to writing a projected volume on plant geography, and the extended treatment he was preparing for his original work had to be drastically curtailed in *The Origin of Species*. Had Darwin carried out his earlier plan to overwhelm his readers with an exhaustive study thousands of pages in length, the course of history would have been far different. As it was, his direct influence on the development of biogeographic concepts became known only after his death and the publication of his correspondence. He discussed biogeographic principles and theories at great length with Gray, Hooker, and Wallace, all of whom made major contributions to the field. In his manuscripts, Darwin anticipated Forbes's work on the distribution of alpine plants and developed his ideas in even greater detail than did Forbes himself, although the credit for this work was lost because Darwin failed to publish.

The excellence of Darwin's biogeography has lately been stressed by Philip Darlington, who remarks that "Eminent or not, as an evolutionary zoogeographer he was extraordinarily, almost incredibly, right about a hundred years ahead of his time." [25] But Darwin was not particularly eminent as a biogeographer, in spite of the fact that he was the first to develop the methodology of the science as it now exists, and in spite of his actual contributions. His work, his methods, and his genius were forgotten by those who came after him. Yet it is not just the fact that Darwin wrote no systematic treatise on biogeography that prevented his gaining full recognition for his accomplishment. Darwin approached both biogeography and geology from the same point of view. And in both these fields his work has been misunderstood for much the same reasons. In conformity with the older notions on the nature of induction, it was formerly assumed that when studying evolutionary biogeography, one should begin by merely describing the geographical distribution of organisms. One would then match the configuration of the continents with the patterns of distribution and make up an evolutionary story consistent with the facts. Anomalies could be explained away by invoking vanished land-bridges or climatic changes. The result would be a picture of distributional changes using evolution to account for the observed relationships. But Darwin did nothing of the sort. For essentially the same reasons that he had embraced geological uniformitarianism, he insisted on interpreting distribution in

terms of the present configuration of the continents: he realized
that scientific progress does not occur without a truly critical
testing of hypotheses. Darwin objected strenuously to the habit
of his contemporaries of throwing up or sinking a land-bridge
whenever a fact could not readily be explained.[26] Such *ad hoc*
hypothesizing was the common practice among biogeographers
of the time, with the conspicuous exceptions of Darwin and
Wallace.[27] If one allows oneself the luxury of land-bridges as
rationalizations—like catastrophes in geology—one can explain
anything. There may well have been quite different connections
between continents in the past, but their existence must be verified in terms of independent evidence, and not invoked merely to
explain away difficulties.

If we examine Darwin's approach to biogeographical problems
in greater detail, we see how it exemplifies his methodology in
general, and how it fits in with the rest of his ideas. One aspect of
his work, a consideration of dispersal mechanism in diverse
groups, has already been mentioned. He first considered how
particular forms might be transported and on this basis predicted
how, if the hypothesized past events had occurred, the organisms
should now be distributed, and then looked to actual distribution
patterns for confirmation. In other words, he constructed a
model involving mechanisms of dispersal. From this approach, we
can easily understand why it was Darwin who first experimented
upon the effects of seawater on seeds. His experiments, reported
in a series of papers, constituted an integral part of the argument
in *The Origin of Species*.[28] He also dealt with such analogous
phenomena as seeds passing undamaged through the guts of birds,
seeds sprouting from mud taken from the bodies of animals, and
clams attaching to the feet of ducks.[29] In *The Origin of Species*
he points out how amphibians, whose eggs cannot withstand the
action of seawater, are invariably absent from the oceanic islands
of the Pacific.[30] In theory at least, it should be possible to infer
the kinds of barriers which have previously existed, by considering the dispersal mechanisms used by the organisms of any given
area. This could be done without any real knowledge of taxonomy, although it would be folly to overlook such useful evidence. And each group, having somewhat different requirements
for success in crossing barriers, would provide evidence relevant
to particular aspects of historical geography.

An exchange of letters between Darwin and Wallace brings out some of the kinds of reasoning each employed and suggests some differences in approach.[31] Wallace held that during the Tertiary period, the faunas of North and South America were far more distinct than they are at present. To this Darwin objected that during the Pliocene, the only period for which adequate data seemed available, certain groups were shared by both areas. There was a real methodological problem at issue, one which is very common in attempts to infer past events from subsequent effects: the direction of change. The problem has already been treated with respect to the inversion of strata in geology, where its solution is straightforward; in the discussion of comparative anatomy, it will reappear in the form of criteria for distinguishing ancestral from derived conditions. Wallace replied to Darwin's objection by showing that, as a general rule, a given group of organisms was particularly numerous on only one of the continents, while on the other continent the group had fewer representatives, and these seemed to have been invaders. Thus the glyptodonts and armadillos were most diverse in South America, while only a few species seemed to have found their way north. On the other hand, elephants, horses, and certain other groups were diverse in North America and closely allied to Eurasian forms, while fewer occurred in South America and could be treated, therefore, as invaders from the north. The principle may be extended to the generalization that a proportionately small percentage of any fauna will be fit as invaders, since any intervening barrier, however slight, will act as a kind of "filter" to at least some of them. Simpson has developed and supplemented this argument, demonstrating that Wallace's interpretation was essentially correct.[32] The precise differences in approach between Darwin and Wallace need some additional study, but it would seem that Darwin tended to concentrate on the effects of different dispersal mechanisms on patterns of distribution, Wallace more on the influence of barriers in restricting faunas conceived of as units. Thus, Wallace was more orientated toward historical explanation for classes of phenomena, Darwin toward reasoning from the effects of the properties of individuals upon the overall pattern of distribution. An extension of this hypothesis will be developed in later chapters to explain their disagreements on a number of other issues.[33]

Darwin's success in biogeography was largely due to the same grasp of the intricacies of applied logic characterizing the rest of his work. He was, for instance, very sensitive to the problem of negative evidence. Biogeography is largely based on explanatory hypotheses which are supported by how well they cast light upon the known facts of distribution. They are rejected, not because some evidence which might support them is not available, but because the facts are better explained by some contrary hypothesis. It was probably for this reason that, in a letter to Hooker, Darwin criticized Humboldt's notions on the migration of plants.[34] Darwin argued that if one were to find that plants common to both Europe and Tierra del Fuego occurred in the intervening area, then it would give one every reason to believe that the plants had migrated between the two areas. But, he thought, a lack of intermediates would be of little significance. The logic behind this conclusion is clear enough. Suppose that there were no such plants in the intermediate region. This could mean *either* that there had been no migration, *or* that the plants had ceased to live along the migration route. Absence of evidence for a migration, therefore, is purely negative and not to be trusted. If it is objected that there is a finite probability that some plants will remain along the migration route, this must be conceded. But how probable is it? One simply does not know: perhaps nine chances in ten, perhaps one in a billion. And unless one has, as perhaps one might have under certain conditions, some means of calculating the probability that such forms will be present, one has no basis for argument. There is no justification for assigning arbitrary probabilities in arguments of this form.[35] A logically equivalent problem arises in a different context with respect to gaps in the fossil record, Darwin's analysis of this problem being one of his major contributions to knowledge. In *The Origin of Species,* Darwin dealt extensively with the lack of fossil evidence for evolution and showed quite effectively that as the fossil record is incomplete, it was no real difficulty for his theory.[36] This conclusion has been borne out by subsequent experience, and well-trained paleontologists are quite sceptical of negative evidence.[37] The same problem occurs time and again in evolutionary biology, and it will be mentioned in relation to a variety of issues.

THE UNITY OF DARWIN'S METHOD

Darwin's biogeography may be interpreted as the consequence of his having applied the same methodology here as elsewhere. Thus, the biogeographic argument for evolution may be fruitfully compared with the first of Darwin's great syntheses, the theory of coral reefs. There are a number of formal similarities between the two theories. Both were discovered by reasoning from theoretical principles and by uniting disparate elements: subsidence and coral growth on the one hand, barriers and dispersal mechanisms on the other. For verification, the parallel is even better. The coral reef theory implies that in certain areas, but not others, particular types of coral formation will occur. The evolutionary theory of biogeography specifies with equal precision the types of faunal and floral peculiarities which should be expected under particular conditions of physical geography. In both theoretical systems, the facts are explained in minute detail; they cannot be explained otherwise, and it is only by ignoring the rationale of the argument that the conclusions are to be rejected.

In a still broader context, it is clear that Darwin's ability to interrelate a wide range of theoretical ideas gives one of the most useful clues to the source of his ideas in general. The evolutionary hypothesis was largely generated by conjoining previously unrelated and quite independent systems of thought. This kind of synthetic approach is basic to Darwin's use of biogeography in a variety of situations. His geology is tested by its implications for biogeography. Biogeographical ideas in turn fit in with evolutionary hypotheses, which ultimately relate to theories encompassing the entire scope of biological knowledge. All these theories are tested against each other through their joint implications and mutual coherence. As Darwin's understanding developed, each experience was incorporated into a foundation upon which he could build anew. Darwin was, above all, a system maker, and his system was a tool both for discovery and for verification.

However true it may be that a new manner of looking at the data of biogeography was necessitated by contradictions in the old theories, the new, evolutionary hypothesis came from outside the conventional system or paradigm.[38] Louis Agassiz, in a work published in 1857, easily drew upon the data of biogeography as illustrative of reason on the part of the Creator.[39] The rich diversity of forms—involving an enormous redundancy in that the different species or even orders often occupied analogous positions in the economy of nature—could support a creationist interpretation through the concept of plenitude.[40] Perhaps God loved variety for its own sake. He must have created as many species as he did in order to impress us with his power and goodness. The situation reappears in morphology: a vestigial organ means not that God created something useless, but rather that he chose to construct all beings according to a single plan or Idea and left in the useless parts for aesthetic purposes. Defective or not, the old morphology and biogeography formed a system which allowed the mind to rest, untroubled by the possibility of a contradiction. Likewise, there was just enough truth in the fact that the constitution of a fauna, or the appearance of a landscape, are determined by ecological factors, that biogeography had no pressing need for a new theoretical foundation. It is clear from Darwin's notes that his confidence in the older theories was shaken when special-creationist preconceptions failed to serve as a useful basis for investigation. And probably such disturbances predisposed him to seek a better explanation. Still, one could always try to rework the creationist ideas. Darwin, having already elaborated a system of historical geology, with all its progressive changes, was well equipped for historical thought, and naturally sought for some kind of reconciliation. The attempt to verify hypotheses about distribution in the light of differences in dispersal mechanisms provided an adequate clue to the evolutionary determination of the observed phenomena. The study of invading populations, and the development of an effective method for reasoning about their activities, led Darwin to conceptualize biogeography according to a point of view most appropriate to the discovery of an evolutionary order. If his previous geological work was not an indispensable element, at least it was of considerable use. Unfortunately, the evidence is not adequate to establish the degree to which Darwin felt he should find some solid grounds for believing that evolution must have occurred.

The fact that it was Darwin, not someone else, who discovered evolutionary biogeography may be explained in part as an accident of rich and fortunate experiences. But one should also remember that Darwin's attitude toward nature differed fundamentally from that which characterized such men of the times as Agassiz. Not content to emote over, and illustrate, the wonders of God's works, he sought out the deliberate and critical development of new systems of hypotheses. Ever seeking for a new model, and incessantly conscious of the implications of old ones, he continued throughout his life to generate new ways of looking at familiar objects. It has been maintained that scientific revolution results from the failures and contradictions of the prevailing system.[41] Such may well be the case for conventional scientists, but Darwin was an exception. He restructured traditional fields and erected new paradigms when the positive development of his ideas suggested something new. The success of at least some revolutionary thinkers may thus be attributed, not to sociological forces, but to an innovative mentality.

3. *Natural Selection*

The theory of natural selection is certainly Darwin's most important contribution to knowledge. It remains, with such implications as evolution, the dominant organizing principle in biology. Natural selection is a remarkably simple idea—so simple that Huxley, for instance, wondered why he hadn't thought of it himself. Organisms differ from one another. They produce more young than the available resources can sustain. Those best suited to survive pass on the expedient properties to their offspring, while inferior forms are eliminated. Subsequent generations therefore are more like the better adapted ancestors, and the result is a gradual modification, or evolution. Thus the cause of evolutionary adaptation is differential reproductive success. Superficially the principle appears to be so straightforward that it should not be difficult to understand.

DARWIN'S PRIORITY

In view of the evident simplicity of natural selection, one may wonder why it was not thought of before. In fact, historical research has unearthed a number of anticipations, both real and imaginary. After the publication of *The Origin of Species* some of these were pointed out. Although Darwin had not been aware of them, he appended a "historical sketch" to later editions of *The Origin of Species*, but retained credit for thinking up the idea himself, for seeing its implications, and for establishing it as a recognized theory supported by factual evidence. Yet Darwin's originality has long been, and continues to be, a bone of contention. In the last century, the novelist Samuel Butler attempted to show, in his *Evolution Old and New*, that Darwin had stolen the

idea of evolution from his predecessors. Darwin never gave Butler the pleasure of a reply.[1] Even natural selection is held not to
be Darwin's original idea. Zirkle has enumerated a number of
examples of what seems to be natural selection, or its elements, in
the works of many writers, among these: Lucretius, Empedocles,
Maupertuis, Buffon, Geoffroy Saint-Hilaire, Naudin, Prichard,
Lyell, Lawrence, and Mathew.[2] More recently, Eiseley has attempted to demonstrate that Darwin took the idea of natural
selection from the writings of Edward Blyth without giving due
credit.[3] Yet de Beer and others have replied that Blyth used
natural selection to refute evolution, and it seems clear, therefore,
that he cannot really have understood it.[4] The attack by the
geneticist C. D. Darlington is largely based upon the notion that
natural selection has been suggested by many earlier writers, and
his evidence is stated in such a fashion as to make a charge of
plagiarism obvious.[5]

A basic approach in all these attacks has been to accuse Darwin
of lacking an awareness of history. The literature is culled for
passages suggesting evolution or natural selection, and these are
easily found. It is then argued that because credit must go to the
worker who first conceived of an idea, the discovery should not
be attributed to Darwin. Yet in a sense it is the critics, not Darwin, who lack historical perspective. One has no trouble finding
other great scientists whose work was anticipated. Copernicus,
for example, was not the first to conceive of a heliocentric universe, but there is every reason to consider him the architect of
the revolution which bears his name.[6] Darwin likewise accomplished a basic reorientation in our view of the universe, whatever may have been the speculations of his predecessors. With
respect to both great revolutionaries, it is clear that their innovations amount to something more than stumbling upon a new idea.
They had not only to generate a different hypothesis, but also to
invent a new way of dealing with a traditional subject matter,
and to demonstrate its advantages by means of a theoretical synthesis. And none of Darwin's supposed forerunners accomplished
anything of the sort.

The arguments over Darwin's priority reflect nothing more
than the difficulty of understanding his theory. Darwin himself
was very much discouraged in his attempts to explain natural
selection to his colleagues, and this may be one of the reasons

why he so long delayed publication.[7] He points out, in his autobiography, that the joint paper of Darwin and Wallace was not adequately understood by the audience, and that it took the long discussion of *The Origin of Species* to win many converts.[8] It would seem that there is something about natural selection, in spite of its apparent simplicity, that made it incomprehensible to the vast majority of biologists—at least until they had considered the matter at length. Even after *The Origin of Species* was published, natural selection did not win immediate or universal approval, nor was it understood as well as one might think. Huxley is generally given much of the credit for popularizing Darwin's theories, yet Huxley himself admitted that *The Origin of Species* was most difficult for him.[9] After a lecture by Huxley at the Royal Institution, Darwin wrote to Hooker in 1860 that "He gave no just idea of Natural Selection." [10] Darwin concludes that "After conversation with others and more mature reflection, I must confess that as an exposition of the doctrine the lecture seems to me an entire failure." [11] One may suspect that even though natural selection was accepted by many of Darwin's contemporaries, they nonetheless failed to fully understand it. Indeed, the concept remains an exceedingly difficult one, even for professional biologists. Reflecting upon the genesis of the theory in the minds of its discoverers may help us to see just what has been the problem.

MALTHUS AND THE DISCOVERY OF NATURAL SELECTION

The parallels in the manner in which both Darwin and Wallace discovered natural selection are as striking as the common elements in their evolutionary biogeographies. Wallace records how, during a malarial attack in February, 1858, he meditated about Malthus's *Essay on Population* and considered the powerful effect of mortality on rapidly increasing populations, in both man and animals. He notes that "while pondering vaguely on this fact there suddenly flashed upon me the *idea* of the survival of the fittest—that the individuals removed by these checks must be on the whole inferior to those that survived." [12] In his autobiography Darwin asserts that he approached the problem of evolutionary mechanisms through the study of artificial selection, but was unable to see how to apply selection to organisms under natural conditions. He says:

> In October 1838, that is, fifteen months after I had begun my systematic enquiry, I happened to read for amusement "Malthus on Population," and being well prepared to appreciate the struggle for existence which everywhere goes on from long-continued observation of the habits of animals and plants, it at once struck me that under these circumstances favourable variations would tend to be preserved and unfavourable ones to be destroyed. The result of this would be the formation of new species.[13]

Both Darwin and Wallace thus would seem to have obtained some crucial insight through reading Malthus, yet what this was has not, up to the present time, been adequately explained. De Beer maintains that Darwin overemphasized the significance of Malthus in his discovery.[14] The same general conclusion is reached by Smith, who asserts that Darwin had his theory fully in mind from the outset of his work on species, arguing that the idea of natural selection was implicit in Darwin's notebooks from 1837 onward, as well as in the writings of Lyell.[15] A number of commentators—there seems almost to be a consensus—support the same general point of view.[16] Yet such interpretations seem a bit odd, in the light of the stress laid upon Malthus by Darwin and Wallace alike.[17] Nor does the charge that, in so many words, Darwin lied about Malthus to conceal his own plagiarism seem to be more than an *ad hoc* hypothesis casting doubt on the argument for plagiarism itself.[18]

The key to an understanding of Malthus's importance has already been suggested by Mayr, who points out that Darwin introduced a new way of thinking.[19] The *Essay* of Malthus's provided the stimulus for conceiving of species in a manner totally different from that which had traditionally prevailed. The innovation may be summed up in a single word: *population*.

A REVOLUTION IN METAPHYSICS

In order to see what a difference the new manner of thinking involved, one must compare it with the system it replaced, an intellectual tradition long dominant in Western

thought. Indeed, a full analysis must extend to the very foundations of Greek philosophy, especially to the influence of Plato and
Aristotle. The group of philosophical ideas that concerns us has
been called *essentialism* by Popper, who has traced the impact of
Plato's metaphysics on political thinking down to recent times.[20]
Even before Plato, Greek philosophy began to experience difficulties in dealing with change. If things grew, or passed away,
they seemed somehow unreal, suggesting that they belonged only
to a world of appearances. Heraclitus, in adopting the notion that
material things are illusory, maintained that all that really exists is
"fire"—that is, process.[21] Plato, on the other hand, elaborated
upon a metaphysical world of changeless reality, of which the
world as we look upon it is but an imperfect picture. His allegory
of the cave, in which all that can be seen are shadows and images,
expresses this conception with the utmost vividness and force.[22]
To Plato, true reality exists in the essence, Idea or *eidos*. The
presumed existence of such a transcendent realm of ideas has
profoundly influenced a variety of historically important schools
of thought, even down to the present day. It accounts for Plato's
adherence to numerology: abstractions of mathematics were envisioned as having more ultimate meaning than mere expedients
in dealing with matter and relationships. Even today, many
mathematically inclined persons are Platonists, and the philoso-
phies of many physicists are saturated with covert numerology.
To Plato, the doctrine of essences had political consequences. His
Republic seeks to justify dictatorship on the grounds of the reality of an ideal state. The earthly republic exists in order to manifest the perfection of the higher reality. Therefore, the citizen's
sole reason for existence is the perpetuation of the state in its
divine form. A despotism is necessary because only the best of men
can understand the transcendent sphere of existence which the
republic must strive to imitate. To anyone who wishes to justify
a dictatorship, Platonic essentialism provides an attractive argument. And Platonism has traditionally been the instrument of
despots, as may be seen in Hegel's symbiosis with the Prussian
monarchy.

In the hands of Aristotle, essentialist metaphysics became
somewhat altered. Aristotle retained the notion of essences, but
held that they do not exist apart from things. His work embraced
concepts of teleology, empiricism, and natural science, making

his writings seem far less alien to a modern biologist than do those of Plato. But if the Platonic *eidos* became less transcendent, it retained a fundamental importance in Aristotelian epistemology. For Aristotle, to understand a thing was to know its essence, or to define it. And definition was not of names, but of essences; concepts were "real," their essences were identified with this reality, and the truth could be reached through precise definition. A true system of knowledge thus became essentially a classification scheme, or an ability to relegate things to explicitly delimited categories.

Plato and Aristotle thus both embraced the notion that ideas or classes are more than just abstractions—that is to say, both advocated forms of "realism." Their metaphysics stands in sharp contrast to nominalism, which asserts that classes are mere conveniences, artifacts of man's thought processes. Thus, to a nominalist, the members of a class do not have the same essence; indeed, the only thing they share is a name. Modern philosophy has tended to be strongly nominalistic. Classification, in most circles, is no longer conceived of as the ultimate form of knowledge. Definition is for the most part looked upon as strictly nominal: definition is of names or symbols, and (insofar as nouns are concerned) the names designate not essences, but things.[23] Such a moderate form of nominalism makes a great deal of sense. When one realizes that words are basically expedients of communication, definition is seen to be important only insofar as it helps us to convey our ideas. Yet definition is so intimately bound up with the discovery of the properties of things, or with the analysis of concepts, that one may readily confound it with quite different processes. It is easy to slip into an Aristotelian attitude and think that when we know the definition of a word we have some kind of insight into matter. Many people are inclined to argue fruitlessly over purely verbal issues: in biology, the dispute over the cellular or acellular nature of protozoa is notorious.[24] A certain dose of nominalism is therefore most desirable in preventing the errors which result from thinking that a definition can be true or false. Yet, carried to an extreme, nominalism can be misleading. It tends to assume, although it need not, that since classes "have no real existence in nature," they must be strictly arbitrary, and that one system of classification is as good as another. And it overlooks the significance of relational properties, so that it may

treat communities or families as if they were just as much abstractions as the class of red books.

Believing as he did that he could define something more than
mere symbols, Aristotle naturally developed a sophisticated philosophy of definition. He advocated hierarchical classification, or
systems of classes within classes. The classes were differentiated
from each other by properties held in common by all members of
each class. A thing was included in the class only if it had the
defining properties. The result of such a method of definition was
a tendency to conceive of things solely in terms of the defining
properties of the class. What was real was the essence and the
differentiae, and the peculiarities of individuals were overlooked.
An implication, of enormous historical importance, was that it
became very difficult to classify things which change, or which
grade into one another, and even to conceive of or to discuss
them. Indeed, the very attempt to reason in terms of essences
almost forces one to ignore everything dynamic or transitory.
One could hardly design a philosophy better suited to predispose
one toward dogmatic reasoning and static concepts. The Darwinian revolution thus depended upon the collapse of the Western
intellectual tradition.

Biological classification was founded in its basically modern
form by Linnaeus (1707–1778), who was an Aristotelian. Many
of the philosophical difficulties of modern biology are due to the
fact that Aristotelian classification and definition have been the
unexamined and automatic practice of working biologists. But a
variety of philosophical attitudes toward classification are possible, and these have led to much controversy. To the Aristotelian,
classes and ultimate reality are one and the same thing. Hence
classification consists in discovering the "real" order in nature
and in expressing, through a system of classes, the properties that
correlate with that order. To the Platonist, the attitude is much
like that of the Aristotelian, except that the classes are seen as
corresponding to an occult order. The goal of classification is to
discover that which is more real than meets the eye—the forms,
eidos, and so forth—and to erect a system based on it. Just as
Plato's politics rested on the notion of an ideal state, Platonic
classification is organized in terms of idealized animals and plants,
and the uniqueness of individual organisms is ignored. For the
same reason that Plato advocated a dictatorship ruled by philoso-

phers, Platonic taxonomy relies upon the authority of the systematist's subjective apprehension of the ideal organic form. To the nominalist, who by definition denies the reality of universals, classes have no real existence, and are merely convenient pigeonholes. Classification, to the more radical nominalist, is completely arbitrary, and one order is as good as another. To the modern biologist, the classes may be looked upon as abstractions, but they are not arbitrary, for they represent an order in nature resulting from evolutionary processes. The classes are not real, but the groups of organisms are. In more philosophical terms, this amounts to admitting the "reality" of relational properties. Classification involves the discovery of relationships, and their integration into a system of theoretical knowledge. Perhaps most important of all, biology has ceased to think in terms of abstract classes or idealized forms such as "the horse" and has turned to considering the interactions between "this horse" and "that horse." We owe this shift in emphasis largely to Darwin.

It should thus be clear that metaphysical preconceptions profoundly influence the course of scientific investigation. The effects of divergent philosophical points of view are readily seen in attitudes toward species. Aristotelian definition leaves no room for changes in properties. In a sense the species *is* the set of properties which distinguish between the individuals of different groups. If species change, they do not exist, for things that change cannot be defined and hence cannot exist. A Platonist, by way of contrast, is interested only in the ideal organism, and ignores the individual differences which are so crucial to an understanding of changes in species by natural selection. If one is a radical nominalist, species cannot exist in principle, and any study of change must treat only individuals. It is possible to accept species as real and still embrace a kind of nominalism, if one looks upon species as individuals. Buffon (1707–1788), for example, would seem to have entertained the notion that a species is a group of interbreeding organisms.[25] This point of view has certain analogies with the biological species definition of the modern biologist: "Species are groups of actually or potentially interbreeding natural populations, which are reproductively isolated from other such groups." [26] A species is thus a particular, or an "individual"—not a biological individual, but a social one. It is not a strictly nominal class—that is, it is not an abstraction or

mere group of similar things—because the biological individuals
stand in relation to the species as parts to a whole. To attain this
divergence in attitude required more than simple affirmation. It
was necessary to conceive of biological groupings in terms of
social interaction and not merely in terms of taxonomic characters.

The new manner of thinking about groups of organisms entailed the concept of a population as an integrated system, existing at a level above that of the biological individual. A population
may be defined as a group of things which interact with one
another. A group of gaseous molecules in a single vessel or the
populace of a country form units of such nature. Families, the
House of Lords, a hive of bees, or a shoal of mussels—in short,
social entities—all constitute populations. The class of red books,
thanatocoenoses, or all hermaphrodites do not. The best test of
whether or not a group forms a population is to ask whether or
not it is possible to affect one member of the group by acting on
another. Removing a worker from a hive of bees, for example,
should influence the number of eggs the queen may lay, and
therefore both worker and queen are parts of the same population. The concept in use is often exceedingly abstract. If one is to
conceive of a population at all, one must use a logic in which the
relationships are not treated in terms of black and white. The
emphasis must be placed on dynamic equilibria and on processes.
The interaction between the parts may be intermittent, disposi-
tional, or even only potential. Entities are components of the
same population to the degree that their interaction is probable.
That the definition of "population" allows for no distinct boundary between populations and nonpopulations creates difficulties
for traditional logic which can only be surmounted with some
effort. And a probabilistic definition of "population," implying as
it does a probabilistic conception of the species, has posed even
more of a problem to the logician. Thus Hull argues as follows:

> The fact that two groups of organisms *cannot* interbreed
> (regardless of the isolating mechanisms) is important only
> in the respect that it follows deductively that they *are not*
> interbreeding. [He continues:] But taxonomists are not
> obliged to predict the future course of evolution. Tax-
> onomists are obliged to classify only those species that
> have evolved given the environment that did pertain, not

to classify all possible species that might have evolved in some possible environment. Until potentially interbreeding organisms actually use this potentiality, it is of only "potential" interest in classification. In evolutionary taxonomy unrealised potentialities don't count.[27]

Hull is here objecting to the kind of vague potentiality talk that proved so misleading and vacuous in the hands of such thinkers as Aristotle. But one needs to draw a line between appropriate and inappropriate contexts, and biologists are fully justified when they employ a terminology loaded with probabilistic implications and expressly designed to convey the idea of potentiality. The probabilistic terminology and manner of thinking that are so characteristic of biology exist for pragmatic reasons and are justified by experience. Their utility is particularly conspicuous in such words as "adaptation" and "environment." [28] The very nature of organic processes demands a probabilistic manner of thinking and a vocabulary adequate to express it. It would be most difficult in embryology, for example, to discuss theory without such terms as "presumptive mesoderm," "regulation," or "potency." And when we are deciding whether or not to eat a wild mushroom, the unrealized potentiality of its poisoning us certainly does count. Just as Aristotelian logic long impeded the progress of biology by necessitating a classification admitting no gradation, modern logic could cope better with systems of thought concerned with probabilistic ideas.

At present, the major source of misunderstanding about the term "population" is a failure to realize the distinction between populations and other kinds of groups. Mayr, for instance, objects to calling a group of asexual organisms a population, although Simpson does not. The divergence of opinion between these two authorities is more than just a disagreement on usage; it reflects differences in attitude of the utmost significance. Simpson uses "population" in an essentialistic or typological sense when he says: "What is classified is always a *population*, defined in the broadest sense as any group of organisms systematically related to each other." [29] At best, this bases "relationship" upon an undefined agreement in taxonomic characters; as stated, the definition is circular. Likewise, Simpson attempts to define "species" in terms of "biological role," which unfortunately does not differentiate between "species" and "subspecies." In Mayr's usage, "population" refers to entities which are more than just nominal classes, and his definition of "species" as particular kinds of populations has the advantage of providing an objective criterion of membership.

GENESIS OF THE THEORY

The basic clue in the discovery of natural selection was the realization that biological groups may form populations, or units of interaction in nature. It had further to be understood how the biological and social entities impinge upon each other in their activities, and how changes in the relationships between the components of the population result from such processes. This in turn had to be related to the problem of adaptation—how it is that properties advantageous to the organisms emerge from social interaction. Conceiving of varieties or species as no more than abstract assemblages, founded on their similarities, precluded any such treatment. Nor could an understanding of evolution be obtained from a consideration of individual organisms apart from their relationships to the breeding community. Darwin attained his new conception of biological reality by a series of innovations, each of which brought him gradually nearer to his ultimate synthesis, and each of which was a departure from traditional thought.

The first step was the analogy between the formation of varieties in domestic productions and among organisms in a state of nature. At this stage it was unnecessary to bring in any consideration of adaptive processes. Darwin was by no means original in realizing that artificial selection may produce changes in the properties of organisms. The comparison was discussed at some length by Lyell, to whom many of Darwin's trains of reasoning may be traced.[30] It was clearly possible for a breeder, by selecting individuals with certain attributes, to form a group characterized by whatever combination of properties he might choose. But this procedure would seem to require an intervention by man for its success; the particular forms selected had to be kept separate from others of their species by an act of volition. And there was no obvious way in such a process to account for the origin of adaptive traits, unless one were to invoke an intelligent agent.

The comparative reproductive success of individuals within the group or population simply had not occurred to Darwin. All that he had understood at this stage was that a distinctive assemblage may be formed by some action which puts like things together. What the action was, except when man intervened, had yet to be discovered. But even this early in Darwin's investigation, one may see a rudiment of population thinking and a key to the manner in which species are formed. In his first notebook on the transmutation of species, Darwin says: "A species as soon as once formed by separation or change in part of country, repugnance to intermarriage—settles it." [31] In the second notebook he expresses the view that varieties may be formed "without picking" —that is, without selection—and that such a change could produce the sterility necessary to keep species distinct.[32] Thus, Darwin had begun to grasp the notion of speciation, the splitting of populations into permanently separated groups. Yet this was only a beginning.

The next step was an understanding of the relationship between environment and adaptation. Darwin approached this question from the perspective of extinction. Early in his first notebook on transmutation of species, he recognized that the cause of extinction is "non-adaptation of circumstances." [33] In the second he refers to the "wars of organic being," in which one species replaces another.[34] We come to a most crucial distinction. One might think that such a picture of well-adapted species replacing the inferior ones was an adequate conception of natural selection. Many students of Darwin have expressed this view and have constructed sweeping treatments of Darwin's work based on it.[35] But they are grievously in error, and their criticisms of Darwin's work, importance, and even morality collapse with the refutation of their basic premise. It is perfectly true that if a group of organisms had some property, the survival of that group would be favored *once the property had been evolved;* but this does not explain how that property might have originated. Certain workers have proposed what has been called "group selection" to account for altruistic behavior on the part of some animals. The survival of the entire population because of an adaptive trait in any of its members is thought by proponents of this hypothesis to be an effective evolutionary mechanism. Nonetheless, it is not the same as the type of natural selection discovered by Darwin and Wallace, and its importance may be challenged on a variety of grounds.[36]

In its effective form, natural selection depends on the differential reproduction of individuals, especially biological individuals. However, social individuals may also be selected, provided that the component organisms of the social unit have much the same hereditary makeup (kinship selection). Thus, the altruistic behavior of mammals toward their young is explicable in terms of the fact that the family, a social entity, favors the differential survival of the genotype for all its members. Likewise, Darwin treated another kind of family, the hive of bees, as an individual society and a unit of reproduction.[37] He refers to just this kind of relationship when he asserts: "The case being, as I believe, that Natural Selection cannot effect what is not good for the individual, including in this term a social community." [38]

To form an accurate theory of natural selection, it was necessary to see how the success or failure of an individual could affect the properties of the species by gradually altering the proportion of individuals with a given characteristic. This required the notions of population dynamics, and extinction was inadequate. For this reason Darwin was fully justified in rejecting the view that his supposed forerunner Naudin understood how selection acts in nature.[39] One may say much the same for a number of alleged precursors of Darwin: they did not appreciate the nature of population phenomena. Nor have some of Darwin's recent critics. For instance, C. D. Darlington criticizes Darwin for expressing uncertainty as to the manner in which the sex ratio has evolved.[40] Darlington invokes group selection: species with a one-to-one ratio of males to females are less likely to become extinct than those with other ratios. But although it is possible to account for the maintenance of an adaptive sex ratio by group selection, it is far less easy to account for its origin by such a mechanism. One might conjecture that it arose quite accidentally, as a consequence of some other process; yet this is not necessary, for recent research shows that the sex ratio is affected by the ordinary kind of natural selection. When there is, say, an excess of males over females, the parents having the largest percentage of female offspring will produce the greatest number of grandchildren, since it will be easier for their children to find mates.[41]

From a consideration of the distinction between "group selection" and the usual kind of natural selection, we may see that Darwin had the necessary components of his theory, but not the theory itself, at a fairly early stage in his investigation. Prior to his reading of Malthus, he was still thinking of species and varieties as mere abstract groupings of things characterized by particular attributes. He had yet to conceive of species as units of interaction composed of biological individuals—as populations, rather than classes. Upon reading Malthus, his attention was drawn to the long-term effects of differences between individuals upon the composition of the population. He ceased to think in terms of idealized forms and became concerned with the activities of individuals. The principle of natural selection was then obvious. Such a shift in emphasis required a fundamental change from the traditional manner of dealing with biological phenomena. It was the beginning of a revolution that transformed not only biology, but man's very conception of the physical universe. It was no longer particularly meaningful to group biological entities merely by enumerating similarities and differences. Knowledge was to be derived from a consideration of how individuals interact with each other, and by the abstraction of the laws governing such activity—laws quite different from those familiar to the science of the times. Darwin never totally escaped from a habit of falling into the old way of thinking, and many times he is inconsistent because of this. Yet he carried out a major portion of the reform himself, and it is this revolution, not the evolutionary one, which most deserves to be called Darwinian.

PHILOSOPHICAL INTERPRETATIONS OF THE THEORY

The significance of Malthus in bridging a conceptual gap in Darwin's thinking is but one of a broader range of topics needing clarification. Let us reconsider the problem of the roots of Darwinism in nineteenth century political philosophy.[42] A widely accepted notion is that natural selection is based on an analogy between the struggle existing in nature and the competition which occurred in society during the Industrial Revolution. Some writers have even gone so far as to argue that the analogy is a false one, and that this undermines the theory of natural selection.[43] Yet this argument is fallacious, for Darwin's theory is not based upon such an analogy at all. That is, no such analogy occurs *as a premise* in his argument. Darwin only used Malthusianism as a heuristic aid—as suggesting one type of mechanism which might be responsible for the observed phenomena. Physicists often do much the same when they look for an inverse-square law, aware that many analogous relationships have been found in the past. The charge of false analogy is an example of psychologism in logic, more specifically of the genetic fallacy. It confounds the historical derivation of the hypothesis with the logical derivation of arguments in support of it. It is also erroneous to argue, from the fact that Darwin makes little reference to Malthus in his notebooks, that the *Essay on Population* was not important in the discovery of natural selection.[44] Malthus provided only a conceptual system or model, not the argument for an empirical proposition, and his contribution is mainly of psychological or historical interest.

From a different viewpoint, namely the philosophical rebellion against essentialism, Darwin's debt to nineteenth century economic and political thought is a very real one. The struggle has deep roots in liberal thought, and it was in the social sciences and in social philosophy that the smashing of ideal forms began in earnest. It is not difficult to see why the attack might begin at this point. The essentialist conceives of the state in terms of a transcendent Idea. The citizen, in his way of thinking, exists only for the purpose of rendering service to society. Any policy which furthers the advantage of the state is desirable, regardless of its effects on individuals. The organization of society is justified on idealistic grounds: the king rules by God's will, some men are slaves owing to the nature of their essences, rights derive from the form of the ideal commonwealth, and so forth. One way to do away with the old order was to undermine the metaphysical arguments which were used to support it, and this entailed a repudiation of the doctrine of essences. A reconsideration of the nature of society, by stressing the importance of the individual, did far more than demolish the philosophical arguments for despotism. Treating social entities not as manifestations of abstract forms, but as the consequence of interactions among individuals, brought about the scientific investigation of society. Human associations, like biological species, are populations, and conceiving of them as such was fundamental to the kind of thinking Darwin owed to his reading of Malthus.

The repudiation of essences has by no means been restricted to the social and biological sciences. It only happens that the crisis was reached there first. The necessity of treating certain aspects of physics on the basis of statistical prediction, such as gave rise to the controversies over Heisenberg's uncertainty principle, is a case in point. The notion that absolutely certain predictions are possible in science seems a bit fantastic to the working biologist, but it has long been a metaphysical ideal among physicists. Chemistry has likewise been confused by deficiencies in theory resulting from typological thought. To the essentialist, the periodic table was a great triumph of classification, allowing the strictest ranking of elements according to their ultimate nature, and permitting a great range of prediction. A gas was noble by definition; therefore one need not look for its compounds. The discovery of xenon tetrafluoride and other "impossible" compounds was therefore delayed. The overthrow of essentialism implies that any classification system or law of nature, be it in physics, biology, or chemistry, is only an approximate generalization. In post-essentialist science, the emphasis has shifted in chemistry toward a greater interest in side reactions and by-products, much as in biology the New Systematics has come to lay greater emphasis on "atypical" variants which the old taxonomy ignored. The nobility of gases is a matter of degree, like inclusion in a biological race. The different attitude toward universals, that was so basic to Darwin's innovation, may therefore be seen as an integral part of a major revolution in Western thought, far more significant than a simple introduction of a historical or dynamic point of view. Darwin's role in this revolution has largely been overlooked by historians and philosophers, and even biologists seem not to have grasped the relevance of his accomplishment to its full scope of implication. The real significance of the Darwinian revolution has scarcely been realized.

The theory of natural selection also involved much that was new in the technique of verification, a topic of great philosophical interest. Much of the difficulty in assimilating the theory arose because to know what constitutes evidence depends on understanding what the theory does and does not imply. Darwin was fully aware how difficult it would have been to demonstrate, through observation of the actual process, the origin of new species, for the process as he knew it obviously took a great deal of time. The problem is no longer with us, since the production of new species in the laboratory, by processes known to occur in nature, is a mere chore.[45] However, direct observations on the process of speciation were by no means necessary; Darwin answered his critics by maintaining that he could verify his theory indirectly by its implication of a large number of readily ascertained facts. Actually, the attack was wholly unfounded and most unphilosophical. In the final analysis, we have no "direct" evidence for the truth of any scientific theory. Everything we know about matter depends on our senses, and we can never come in direct contact with even the most mundane thing. Everyone should know this; but it is sobering to realize that the issue, in one form or another, continues to be argued in scholarly publications.[46]

But leaving irrelevant issues aside, there is a real problem of what facts can be invoked in support of Darwin's theory. Natural selection accounts for certain phenomena of organic nature as resulting from quite orderly processes. As a consequence of the underlying mechanisms, certain historical events are likely to have happened, but the occurrence of others would be improbable. The actual occurrence of such events may be established by studying their effects on organisms, either recent or fossil. The same applies to the main competing contrary hypotheses: Lamarckianism and special creation. Each of the three implies that there will be particular kinds of properties in organisms, and the relative merits of the theories may be evaluated by how well such predictions are borne out by experience.

Arguments in support of any hypothesis, including an evolutionary one, must be based on critical tests. It is altogether simple, yet delusive, to seek in our observations a number of facts which are merely consistent with the theory. As Popper has put it, "Every serious test of a theory is an attempt to refute it." [47] The best one can hope to do is to subject the hypothesis to the broadest possible range of tests; but in so doing, certainty is never obtained. As has been mentioned above, there are certain so-called theories which cannot possibly be refuted and which should, therefore, be regarded as pseudoscientific. The theory of Lamarck is at least to some extent a scientific one, and it has been refuted. This refutation is partly due to Darwin's argument that since conscious striving, presupposed in Lamarck's theory, does not occur in plants, the theory is inadequate to explain all evolutionary phenomena.[48] The theory of special creation, insofar as it is a scientific hypothesis, was also refuted by Darwin, on the grounds that an intelligent Creator should have distributed organisms, not following their taxonomy, but according to their physiological needs.

It seems almost unreal that, among all the theories of science, the Darwinian theory of natural selection has been singled out as incapable of refutation.[49] As has already been suggested with respect to the coral reef theory, this charge rests on a fundamental misinterpretation of the falsification principle. A theory is refutable, hence scientific, if it is possible to give *even one* conceivable state of affairs incompatible with its truth. Such conditions were specified by Darwin himself, who observed that the existence of an organ in one species, solely "for" the benefit of another species, would be totally destructive of his theory.[50] That such an adaptation has never been found is a most compelling argument for natural selection. Additional examples of the manner in which the theory of natural selection may be tested, and further statements of possible conditions of refutation, will be treated in the context of particular tests.

Darwin's argument for natural selection, like the coral reef and evolutionary biogeographic theories, was based upon the construction of hypothetico-deductive systems. That Darwin so conceived of his evolutionary theories is clear in the following quotation from his notebooks: "The line of argument often pursued throughout my theory is to establish a point as a probability by induction, & to apply it as hypothesis to other points, & see whether it will solve them." [51] The argument rests upon a system of theoretical premises: if certain laws were operative and conditions did in fact eventuate, it follows logically that particular consequences may be observed in nature. The agreement or lack of it between predictions and observations is looked upon as a test of the system. I say "a test of the system" because it is not simply the facts or theories in isolation which are tested. The theory is so constructed that the hypothesis describes relationships which form a unit. Hence such terms as "favored races" or "natural selection" are unintelligible in isolation.

We may analyze Darwin's hypothetical language by considering such words as "fittest" in Spencer's "survival of the fittest," or "favored" in "the preservation of favored races in the struggle for life." These terms are in many ways comparable to "adaptation" and "fitness," terms having various and often confused meanings in biology. "Fitness" or "adaptation" do more than merely state the obvious fact that organisms must have certain properties to survive.[52] Such oversimplifications have been argued on the grounds that the only way to know whether or not an animal is fit is by whether or not it survives. The fallacy is that the "adaptation" of an organism means, in actual usage, not the fact of survival, but its probablility. It refers to something analogous to "life expectancy" in the insurance business, and to continue the analogy, it does not prophesy that a given individual will expire on a given date, but only provides a statistically meaningful approximation. It is perhaps largely because this frame of reference has been misunderstood that some writers have asserted that "adaptation," "variation," and "environment" are circularly defined.[53] Certainly in *The Origin of Species*, as in modern biology, such terms are not defined with reference to the differential survival which gives rise to evolutionary changes. Fitness is a physiological concept, in many ways comparable to the efficacy of a tool in accomplishing some particular task. It assumes that different types of structures can accomplish various vital functions with different degrees of effectiveness. Thus, the structure of an eye is considered adapted to seeing, insofar as the eye can form a distinct and detailed image, and the question of its fitness becomes a problem of optical engineering. Some confusion has arisen because the standard of efficacy need not be the same for every context. An eye which works well in water is not necessarily equally effective in air. And since the conditions of existence for organisms are most complicated, it is even argued that "adaptation" should be abandoned because it is not knowable with absolute certainty. An obvious reply would be that one need not expect that much of a word. The term is meaningful within limits that allow one to communicate significant ideas, and no more need be asked.

Further insight into what "fitness" means in the theory of selection may be obtained by considering its functions in the hypothetical system. The theory supposes that there are variations, and this is an obvious fact. It also states that for a given structure, sometimes, two different variants have different effects in furthering the lives of the individuals. Thus it is predicted that *if* there are variations, *if* these are inherited, *if* one variant is more suited to some task than another, and *if* the success in accomplishing that task affects the ability of the organisms to survive in whatever happens to be their environment, *then* natural selection will produce an evolutionary change. Such conditional statements are basic to the Darwinian theory, and this conditionality has given rise to much confusion. None of the conditions are more than crudely predictable in the case of individual evolutionary events. Yet this does not affect the validity of arguments showing that certain facts are inexplicable unless the theory is true and the conditions have been met. The theory need not predict when the proper conditions will occur for evolution, but only what will happen when they do occur. A favored race is, by definition, one which has the efficacious properties, those which *would be* favored by selection if all the other conditions should be met.

If the assumptions basic to the theory of natural selection hold true, and if some unknown factors do not intervene, it follows deductively that evolution must occur. Because of the manifest rigor of this argument, it has been asserted that "natural selection" is a tautology, and the same is occasionally said of its elements, such as the "struggle for existence." The opinion that a tautology is involved has been used to justify some curious notions. Antievolutionists may hold that this renders natural selection irrefutable, hence unscientific. Zealous partisans of the theory, conversely, may believe that the tautologous nature confers some unusual degree of certainty. The same fallacious argument is given by certain philosophical physicists and others in what has been called *conventionalism;* they hold that the a priori certainty of the mathematics involved in scientific thought makes scientific inferences as reliable as mathematical reasoning.[54] The trouble is that in theories of physics and evolution alike, the only tautologies are in the deductive core of the argument. The truth of the hypothesis is justified not on the basis of the tautologous element, but only on the ability of the system to generate true predictions about the material universe. It is logically possible that one of the premises is false. Or perhaps some unaccounted-for element counteracts the effects of selection; indeed, just such a possibility was raised in the form of "blending inheritance." Whether or not natural selection is the cause of evolutionary phenomena can only be determined by experiential testing. The assertion that the theory as a whole is as tautologous as its elements constitutes an outrageous instance of the fallacy of composition: that is, it assumes that the properties of the part are the same as the properties of the whole.

Like Darwin's coral reef and evolutionary biogeographic theories, the hypothesis of natural selection is predictive, testable, and capable of refutation. But in its implications it has no need to predict, with great accuracy, the long-term course of historical events, any more than the law of gravity should be expected to predict the path of every drop of rain. Its implications are statistical, and in this the logic of its verification is in many ways comparable to that of the second law of thermodynamics.[55] Or one could draw analogies with Le Chatelier's principle and the laws of diffusion. The manner of verification of theories such as natural selection, which allow one to make statistical predictions of events, does not differ in principle from experiential testing generally. The difference with respect to historical sciences lies in the degree of complexity of the generalizations and in the conditional reasoning which is rather unfamiliar to most experimentalists. Natural selection does not prophesy—it only implies that matter will tend to assume certain configurations more often than others, and that such regularity is intelligible in terms of theory.

The implications of natural selection are by no means restricted to the brute fact of evolution. Indeed, natural selection does not imply that evolution will occur at all; it only implies that when evolution does occur, it will proceed according to certain rules. The same is true of laws of nature in general. Again, the law of gravity does not predict that an object will fall. For this reason, any objection to the theory of evolution as an explanatory hypothesis on the grounds that it does not predict some particular event may be reduced to an absurdity. The theory does have numerous implications which compel us to accept natural selection as the cause of evolution, and many of these were worked out by Darwin himself. Among the most important is sexual selection, based on the idea that evolution results from differential success in reproduction. The competition for mates *could* result in maladaptive developments quite inconsistent with other hypotheses, but readily explicable in terms of selection. The selection theory, based as it was on population thinking, gave rise to many misunderstandings. More recent developments derived from the same kind of model include the discovery of how population size affects the course of evolution. In a small population, fortuitous events are far more important than in a large one. The theory of probability tells us that the likelihood of getting a continuous series of "heads" decreases with the number of times the coin is flipped. For analogous reasons, increasingly small populations are more likely to undergo large fluctuations in the ratios of genes, owing to purely random occurrences. The larger the population, the more the effects of "sampling error" tend to cancel one another out. Phenomena resulting from chance fluctuations in gene frequency are particularly noteworthy in some circumstances. For instance, on isolated islands, where colonizing organisms arrive in very small numbers, the inhabitants differ from the ancestral population because it is unlikely that the small group of colonizers will have the same gene frequencies that existed in the population which gave rise to it. This implication of selection theory, the "founder principle" of Mayr, does not predict the frequency of any particular gene.[56] It does predict, however, that certain kinds of variation will correlate with patterns of geographical distribution. This kind of prediction has a wealth of analogies in the physical sciences. The kinetic theory of gases, which is also a population concept, implies that minute particles of smoke seen under the microscope will move about irregularly because they are struck unequally by the random movement of gaseous molecules (Brownian movement). With particles much larger than smoke, the effects of individual molecules cancel each other out, and the effect is like that of a steady pressure. The synthetic theory of evolution explains the founder principle in much the same way that the kinetic theory explains the Brownian movement.

In the sense that natural selection makes statistical predictions, it is consistent with the covering-law model of scientific explanation. It might be objected that in many cases (such as the founder principle) this would entail the "prediction" of past events, disagreeing with conventional usage.[57] Again, some even go so far as to insist on the use of terms like "retrodiction" and "postdiction." [58] The former view is a reasonable objection to an equivocation, but the latter is merely an instance of psychologism in logic. On the one hand, it is a seemingly odd, but still valid, usage to employ the verb "to predict" in the sense of following logically rather than temporally. On the other hand, one does "predict" in the temporal sense when one predicts what one will observe when, for example, he studies the present distribution of a group of organisms in order to test a hypothesis about their past history.

Another class of objections to natural selection has roots in metaphysical attitudes toward laws of nature. Platonists, including many so-called philosophers of science, look upon laws of nature as occult forces. This may be true even though they claim to have eliminated metaphysics from their own thinking. Although the generality of physicists have come to regard, say, the force of gravity as anything but an occult force existing apart from matter, a remnant of such thinking still has much influence. This is to be seen in the thinking of those who hold that purely statistical laws are somehow inferior, or that they are underlaid by more basic principles. It becomes even more evident among those who maintain that history is not a science because it has no "laws" like those of physics. This view is justified insofar as "history" is taken to mean an approach stressing the occurrence of individual events. In the sense that there are no historical entities comparable to organisms and chemicals, there is nothing to be ordered according to laws. If by "history" one means an approach, it is no more a science than "experimentation" or "comparison." But it does not follow that "historical" sciences are less "scientific" than "experimental" ones. Once again, we see how metaphysics— covert, yet very real—arises in what one would think is a purely scientific issue.

A most useful analytical instrument for understanding evolutionary theory is an analogy, first developed by Charles Sanders Peirce, between what happens in natural selection and in card games.[59] The comparison is most useful, for not only does it treat population phenomena in terms of everyday experience, but to elaborate it may help to take the metaphysical pathos out of empirical generalizations. In a card game with a limited number of players and limited funds, and where the game is continued long enough, one player after another will go broke, and someone will ultimately win everything. There is nothing mysterious about this—it is all intelligible in terms of elementary probability theory and the rules of the game. Can one call the tendency for the number of players to decrease a law of nature? Does an understanding of the structure of the game allow one to predict and explain the course of events? Or does it seem more reasonable to redefine "law," "predict," and "explain" for the sole purpose of rationalizing one's metaphysical ideas? We may see that there is the strictest parallel between such sampling effects as genetic drift and the founder principle on the one hand, and the decreased probability of going broke with more funds on the other. And in neither the card game nor the history of life on earth does a failure to predict every individual event affect our ability to invoke the principles in question to explain why one or another player goes broke, or why some species become extinct.

The card-game analogy may be extended to cover a number of further criticisms of Darwin. It might be asserted that natural selection explains evolution as a result of chance. What on earth would this mean? Does someone win all the money in a card game by chance? In the sense that a given player wins all the money, the answer is yes; in the sense that such an event should happen, no. The parallel with natural selection is only too obvious. The concept of natural selection seems unsatisfactory simply because it is intelligible by reason. It is no more impressive than the fact that the more fit players—those who in poker, for example, usually fold on a low pair—tend to accumulate money. One wonders how long it will be, if ever, before laws of nature lose their metaphysical pathos, and are looked upon as no more impressive than sound advice on how to win at cards. Is the law of universal gravitation any more profound or metaphysically significant than "never draw to an inside straight"? It seems odd that scientists should revere a thing simply because they do not understand it. But scientists are men like anybody else, and superstition is one of the commonest attributes of our species. Human nature seems all but universally attracted toward an occult force of gravity. Likewise, some feel that some innate tendency toward perfection should be invoked to account for evolutionary progress. We might just as well explain our losses at the gaming table as due to some Impecuniary Influence beyond our control.

The card-game analogy has particular relevance to that most

confused notion, biological progress. In the hypothetical card game, one player after another goes broke, and one of them ultimately accumulates all of the money. One could say that in evolution there is an analogous progression in the improvement of organic structure, although the comparison has its limitations. In the first place, the model of the game does not imply that any particular individual will succeed or fail—indeed, for each player, the chances are against his ending up as winner. The comparison here holds up quite well with respect to evolution; a few species progress (in some sense), but for most the result seems far less desirable: at worst, extinction. This point was well appreciated by Darwin, who held most consistently that progress is not inevitable and that evolution often leads to degeneration.[60]

The inevitability of progress in uncontrolled competition has been called "social Darwinism," yet Darwin embraced no such notions. The only sense in which Darwin believed in progress was that one should expect it to occur in some instances. Progress is explicable as a consequence of the theory, but its individual instances cannot necessarily be predicted. All one may expect is that a few forms should become changed in the direction of increasing functional efficiency. Even to this limited extent, he who uses the concept of perfection in organic nature runs grave risk of deluding himself. Darwin was well aware of the difficulties; it is said that he made a memorandum in his copy of a pre-Darwinian evolutionary tract, *Vestiges of the Natural History of Creation,* saying "Never use the word 'higher' and 'lower.' "[61] One source of confusion is that what constitutes adaptation in one environment is clearly deleterious in another. Thus, although we tend to consider warm blood superior to cold, frogs may be at a competitive advantage over birds and mammals where food is scarce, since warm-blooded organisms dissipate a large portion of the available energy as heat, preventing them from fasting for extensive periods. More importantly, it is no longer reasonable to believe that organisms form a single series from the simple and inefficient to the complicated and dominant. Such thinking is a vestige of the Aristotelian and Neoplatonic notion of a "chain of being," the history of which has been admirably treated by Lovejoy.[62] Modern biology has found it necessary to conceive of organisms as corresponding to many series, none of which is strictly comparable to the others. Different lineages have changed in different ways, and diverse organ systems have become altered at different rates. One may still conceive of adaptation, perfection, or whatever one wishes to call it, in terms of an engineer's criterion of efficiency, but even this has its limitations. One hardly knows where to begin, in ranking such widely divergent organisms as butterflies and orchids. The perfection, even in the sense of engineering criteria, differs from one structure to another. Although man has, relative to a bird, a "better" capacity for reasoning, and perhaps a less wasteful manner of rearing his young, he is distinctly inferior in the structure of his lungs and a number of other organs. But to complicate the problem, one must observe that birds "need" especially effective lungs because of the demands for oxygen imposed by flying. It would be better to compare birds with bats; but then again, the habits of life are not comparable in all respects. In other words, the "perfection" of an organism—its "highness" or "lowliness"—can only mean the summation of the "perfection" of individual parts. But since the individual parts are not comparable to each other, an attempt to create a calculus or scale of perfection must be founded upon the assumption that perfection in one structure is equivalent to perfection in another. But this is absurd, because the mechanical perfection of an eye must be measured on a scale appropriate to eyes alone, and their qualities are not comparable to those of nerves or teeth.

The idea of an absolute scale for "highness" or "lowliness" is an example of essentialist metaphysics which is very common in our everyday thought, and which has profound social implications. The position of an organism in the scale of being is determined by a purely arbitrary or subjective decision as to the relative merits of different organ systems. Usually, the standard has been an anthropomorphic one—man is the ideal by definition—but there is no reason for rejecting any other criterion. Every society has its own arbitrary and subjective standards. Social worth is defined in egocentric terms—"inferior" nations or races are valued in proportion to their degree of resemblance to Plato or Hitler. In modern society, the wholly arbitrary scale is perhaps more oppressive than the egocentric, for it can exist behind a façade of objectivity which yields only to an attack at its Platonic foundations. We are inclined to believe in our ability to rank individual human beings in a scale of health, virtue, or intelli-

gence, and to think that the summation of such orderings allows a
measure of individual worth. A man's health is proportional to
the quality of his organ systems. But which is more important,
sound teeth or sound eyes? There can be no truly objective
standard of comparison, ever. The same holds for I.Q. tests and
grade-point averages, which reflect summations of several different talents, none of which may truly be compared with another.
The result is that the individual excellences of unique personali-
ties are ignored to everybody's detriment: the classic example
being Darwin, who achieved no distinction whatever in formal
education. Although men possess talents in degrees, they possess
no degree of talent. Perhaps the equality of man is a mere ideal
for a just society, but the inequality of men is neither an ideal nor
an objective fact.

Among the most pernicious effects of essentialism is the treatment of laws of nature as if they were the fiats of an omnipotent
legislator—as if, like the laws of man, they derived their existence
from some coercive power. The notion of "selection pressure"
is particularly apt to imply that if organisms do not change, there
must be some inertia or counteracting force to balance its effects.
The results of such confusion are to be seen in the charge made
very early against Darwin that if natural selection acts to produce
evolutionary change, one would expect all organisms to have
been acted upon by it and hence to have attained a "perfection"
equal to man's. But such views overlook the logical structure of
the theory. *If* there is variation, *if* the variants differ in fitness,
and *if* the variations are inherited, *then* there is evolutionary
change—sometimes. When one such necessary condition is not
met, evolution does not occur. Animals or plants may remain at a
lowly grade of organization in particular characteristics simply
because they have not varied so as to produce the more efficacious condition. To attribute a change to selective advantage on
the grounds that this explains the facts is a valid induction. But to
attribute a lack of change to a lack of advantage overlooks the
alternative of no variation, and is, therefore, fallacious (confusion
of necessity with sufficiency). This particular kind of mistake is
widespread, even in authoritative places; for example: "The production of homoiothermy, or of a large fore-brain in the pri-
mates, was probably a major improvement in organization, applicable to a great variety of particular forms living very diverse
lives. However, the problem then arises, why have not all animals

produced these improvements? This can only mean that there are many circumstances in which they are not applicable." [63]

A use of the same kind of fallacious reasoning also underlies one of the major arguments against adaptation: the notion that an ill-adapted form cannot exist because it would have undergone selection to adjust it in conformity with new circumstances. Clearly, an organism may become as well adapted to its environment as the existence of advantageous, inherited variations will allow. Yet nobody would affirm that an ideal variation must inevitably occur. Still, Williams has lately written an entire book largely to dethrone the concept of adaptation in modern biology, and one which frequently wanders into the same old pitfalls.[64] Thus, he erects a rather intricate argument to refute the view that man's intellectual powers arose through natural selection, maintaining that he sees little value for a high degree of reasoning ability in primitive ancestors of men. He speculates that perhaps intelligence is selected among children, and that growth continuing into later stages produced the more extreme development of the brain. He supports this view with the idea that since leadership is largely exercised by the male, one would expect intellectual prowess to have developed in men alone. Yet even leaving aside the questionable premise that intellectual prowess is of no use to women (when they are finding mates or raising children, for example), one must flatly reject the notion that differential selection between the sexes would necessarily lead to sexual dimorphism. A difference in mentality between men and women, even if intellectual ability were advantageous for men alone, is not to be expected unless it were advantageous for women to be stupid, and even then it need not develop. A difference in mental ability could only evolve if, somehow, the variations which produced advancement in brain function were limited to one sex.[65] Even if a particular mental quality were advantageous for males alone, the genes could increase in the population as a whole and be expressed in both males and females. If such genes were disadvantageous to women, a modifier gene expressed only in females could produce a dimorphism—but only if such a variation occurred, and there is no reason to think that it must occur. To be sure, the question of determining the manner of origin of adaptive traits is difficult; but as we will see in later chapters, the necessary methodology was long ago worked out by Darwin.

The human mind habitually inclines toward those ideas which

present themselves most vividly to the imagination, while those
less readily conceived seem not convincing. It is no wonder,
therefore, that natural selection, which is particularly difficult to
picture in the mind's eye, is misconstrued by those who conceptualize it through inappropriate constructs. The metaphorical
"struggle for survival" is a case in point. In the Malthusian economics, men engage in a bitter conflict for those resources which
are in inadequate supply. Similarly, in the title of *The Origin of
Species* Darwin refers to "the preservation of favoured races in
the struggle for life." It is easy to imagine that the essential
element here is struggle or conflict. It seems a perfectly natural
inference that Darwin owed to Malthus no more than the notion
of a "struggle" and that this struggle is the efficient cause of
evolution. We are easily led to conjure up a vision of suffering
and greed. It seems—if not repugnant to the sensibility and perhaps unsatisfying to the imagination—hardly conceivable that
conflict should act upon organisms to produce evolutionary
change, and such an objection has frequently been raised against
Darwin's theory. But the whole picture is in error. Struggle is not
the basic element in natural selection, but only a description of
the conditions under which natural selection does in fact occur.
"Natural selection" designates not the struggle, but the preservation of favored races accompanying that struggle. The common
feature of conflict in both Darwinian evolution and Malthusian
economics has nothing to do with the essential element, namely
the interaction of individuals in a population. Natural selection is
differential reproduction with its causes, nothing more.

Since struggle is readily pictured in the imagination, then, it is
easily misconceived as a causative agent.[66] Natural selection,
however, is not so easily conceived. How, for example, can it
"act" on an organism? One might think that there must be some
agent or force to be identified with the name. But this would be
absurd, for natural selection is not a force or a tangible object,
and it does not act upon organisms like a potter's hands molding
clay. The term simply describes what happens, when certain conditions are met, to the relationships between the component units
of a population. The situation is in many ways analogous to the
confusion which resulted when men began to conceive of action
by gravity at a distance, or of the undulatory theory of light, and
postulated vortices and ether to account for what seemed un-

imaginable. In biology, the same commonsense attitude has led to
the denial of competition and adaptation, which likewise are not
tangible objects, intrinsic properties, or concrete forces.[67]
Among intellectual historians, we have the spectacle of various
writers accusing Darwin of hypocrisy for his perfectly innocent
statements that he learned something when he read Malthus, on
the grounds that he was already aware of struggle in nature.[68]
They have simply missed the point. But what can be done to
avoid such errors in the future? The answer, as I see it, is to
abandon the study of words and to derive our understanding
from concepts. The structure of Darwin's systems explains his
success and failure alike. When the process through which his
discovery was generated has been understood, there is no reason
whatever to treat his perfectly ingenuous accounts of the discovery as mistaken, contradictory, or hypocritical. Our reconstruction, or Darwin's reconstruction, may be verified when we see
that certain insights follow from his premises, while others do
not; the former he grasped readily, the latter could only be obtained, if at all, by restructuring his thought, or by taking another approach. When we understand Darwin's system it is
easier to see, for example, why he experienced difficulties in explaining the causes of evolutionary divergences, as recounted in
his autobiography:

> But at that time I overlooked one problem of great im-
> portance; and it is astonishing to me, except on the prin-
> ciple of Columbus and his egg, how I could have over-
> looked it and its solution. The problem is the tendency in
> organic beings descended from the same stock to diverge
> in character as they become modified. That they have di-
> verged greatly is obvious from the manner in which spe-
> cies of all kinds can be classed under genera, genera under
> families, families under sub-orders, and so forth; and I can
> remember the very spot in the road, whilst in my car-
> riage, when to my joy the solution occurred to me. . . .
> The solution, as I believe, is that the modified offspring
> of all dominant and increasing forms tend to become
> adapted to many and highly diversified places in the econ-
> omy of nature.[69]

This simple oversight may seem odd to us as well as to Darwin,
and at least one commentator suggests that Darwin really had the

answer all along.[70] But in view of the theoretical approach used
by Darwin, his missing this particular relationship is hardly surprising. There is nothing in his model about the diversity of
niches in nature, and it was not until he began to work out the
implications of his theory for ecology that the problem, or its
answer, occurred to him. Any working scientist should be familiar with the same kind of experience. Such oversights hardly
reflect on the investigator's talents; rather they should be looked
upon as manifesting those neurophysiological laws that may serve
as a basis for a scientific approach to history.

INNOVATION AND METHOD

The theory of natural selection, now looked upon as the
foundation of evolutionary biology, was the third of Darwin's
great syntheses. Darwin deserves full credit for this innovation, as
it was he who first conceived of natural selection in terms of a
developed, hypothetico-deductive system, capable of explanation,
prediction, and experiential testing. All of his so-called forerunners, without exception, either only conceived of elements in the
theory, or did not understand its significance. Wallace alone deserves credit for his independent discovery. Darwin and Wallace
merit particular respect for having developed the theory of natural selection through a process of "retroduction": that is, they
were aware of a phenomenon, and successfully sought out an
explanation in superficially unconnected processes.[71] The
method through which this insight was obtained would seem to
have been orderly and rational. Further evidence for this thesis
emerges when we compare the theory of natural selection with
Darwin's other major syntheses. All three of Darwin's theories—
coral reefs, evolutionary biogeography, and natural selection—
invoke a type of order resulting when processes and conditions,
random in relation to each other, interact: subsidence and coral
growth, barriers and dispersal, variation and fitness. All three are
based on abstract models which have numerous implications and
can readily be tested. There is a clear-cut series of rational operations leading to each theory, although natural selection differs in
having an added difficulty of shifting from groups to populations.
The development of thought from invasions to evolution, and
from artificial selection and extinctions to natural selection, ac-

companied by a shift in theoretical perspective, explains the facts of the discovery which have long seemed contradictory. All that we need yield to the notion of intuition is a grasp of the structure of models, an ability to vary them (perhaps at random), and attention to that moment when their implications are significant. Seeing in Malthus how the interaction of individuals in the same species may be affected by the intrinsic properties of each organism, and how there could be cumulative effects, Darwin and Wallace were able to conjoin all the disparate elements into a unitary system which constituted the theory of natural selection. The discovery was, above all else, a triumph of reason. If banishing intuition from our conception of the process of discovery deprives us of a sense of mystery, it nonetheless permits us to analyze that process in a far more satisfying manner than did the mythological accounts.

4. Taxonomy

The theory of evolution had an enormous impact upon systematics. For although it is possible to erect classification systems without assuming that there is any particular cause for the similarities and differences between the entities which are classified, an awareness of the underlying mechanisms cannot fail to affect our thinking about the goals and significance of taxonomy in general. Darwin's own views on the relationships between evolution and classification have been but imperfectly understood. He devotes a separate section to such problems in *The Origin of Species*, and often-quoted passages occur scattered about elsewhere in the same work and in *The Descent of Man*.[1] Occasional references to taxonomic theory are found in many of his other biological writings and in his letters and notebooks. One of his botanical works includes a chapter primarily concerned with strictly philosophical aspects of classification.[2] Perhaps the most useful evidence relevant to Darwin's views has hardly received any attention at all: his taxonomic writings, particularly the *Monograph on the Sub-class Cirripedia*, works which are but rarely read, especially from this point of view. The problem of deciphering the primary sources is particularly difficult, not only because, like many systematists, he often stated his conclusions without a full exposition of his arguments, but also because his work on barnacles was written before he published his evolutionary theory. Therefore his methods and philosophy can only be inferred from indirect evidence and oblique statements. However, if Darwin had a coherent philosophy of classification, it should be possible to draw together his numerous and scattered statements on these matters, and to show that they are mutually consistent. It should be possible to test the validity of any such hypothetical interpretation by showing how he applied the various concepts in practice. Our need for historical analysis of Darwin's taxonomic theory has become crucial, because his actual opinions have been obscured through partisan, and sometimes biased, appeals to his authority.

The problem of understanding Darwin's views and methods in taxonomy is no simple question of historical fact. Taxonomy is a highly controversial subject, and the issues are inextricably bound up with philosophical disputes which have endured for centuries. The problems are so important that no biologist can totally avoid facing them. They are so controversial that objectivity in their study is perhaps an unattainable goal. And the issues, both biological and philosophical, are often so recondite that an unequivocal solution seems impossible. Nonetheless, some grasp of them is necessary if we are not to overlook a major portion of Darwin's thought. Therefore an explanation of Darwin's views on the philosophy of classification is attempted in the present chapter. It has been essential to interpret the data in terms of my own theoretical point of view, one which by no means represents a consensus of authoritative opinion.[3] An analysis is given which seeks to bring out significant distinctions without which Darwin's views cannot be understood. This will first involve a consideration of the concept of natural classification, which is relatively simple and for which there is little difficulty in ascertaining Darwin's views. There follows a discussion of a related issue, that dismal morass of verbal confusion called the "species problem." Once these issues are clarified, it will be possible to demonstrate how Darwin applied his theoretical concepts to concrete problems.

EVOLUTION AND NATURAL CLASSIFICATION

The necessity of classification systems has long been recognized. Indeed, for the very communication of general ideas, it is essential to group things together. Primitive man's vocabulary of terms for plants and animals is often rich and varied, reflecting his concern for a wide diversity of organisms. For learned discourse, a far more elaborate and precise terminology has been developed. Yet a coherent and useful method for ordering and naming groups of organic beings was not developed until the middle of the eighteenth century, when Linnaeus applied the so-called Linnaean hierarchy to this end. The Linnaean hierarchy, taken over from Scholastic logic, involves the erection of classes within classes, so that the class of vertebrates is divisible into subgroupings of birds, mammals, and so forth, each of which in turn includes smaller groups, so that all of nature is encompassed by a single system. Linnaeus himself differentiated between "natural" and "artificial" systems, the latter being mere conveniences—arbitrary assemblages, like the class of "tall men." A natural classification system, on the other hand, had some deeper meaning. As Darwin puts it: "This classification is evidently not arbitrary like the grouping of the stars in constellations." [4] Such a system thus refers to something which we do not construct, but rather discover. Nor are the classes simply collections of things which have happened to strike the eye as similar.

But affirming a distinction between arbitrary and nonarbitrary groupings has philosophical consequences. A nominalist would reject such a distinction: to him, a class is always an artifice created by men for the purpose of convenience. This view is sound as far as it goes, but it overlooks the distinction between classes and the order upon which they are founded. The term "natural system" has been used in at least two different senses. "Natural system" may mean, on the one hand, the order which is taken as a standard of classification; on the other hand, it may mean the system of classes which expresses this kind of order. And the distinction holds, whether or not we are deluded in thinking that such an order exists, and in spite of how many kinds of classification systems may be based upon that order. Thus a natural system is based on an order which is "real," in the sense of "having real existence in nature," even though extreme nominalists would contend that this is impossible. "Real" and "natural" on the one hand, and "artificial" and "non-natural" deal with the same kind of distinction. But one's interpretation of the meaning of such terms as "real" differs with one's metaphysics.

To the Platonist, what is "real" is the *eidos*, Idea, or archetype. Hence, "natural system" is identified, not with an order in nature, nor with a system of classes corresponding to it, but with the transcendent Ideas which to him are more real than the organisms themselves. What matters to the Platonist is the concept. Platonists insist that "natural system" is not based on phylogeny or on evolution because our knowledge of phylogeny is but an imperfect picture of a logically prior, intuitively apprehended order.[5] To the Platonist, definition is of Ideas, and taxonomic groupings are the names, not of organisms, but of concepts. Hence, those biologists who embrace Platonism quite generally believe that genera or species are abstractions rather than the things which their names designate. Similarly, the transcendentalist Louis Agassiz (1807–1873) maintained that species are "categories of thought," a view which Darwin naturally dismissed as empty verbiage.[6]

As Mayr has repeatedly pointed out, an essentialist approach to taxonomy was a major barrier to the discovery of evolution and natural selection.[7] To Agassiz, again, evolution was not possible, because categories of thought would have to be unchanging and eternal. Some forms of Platonism, however, are not opposed to evolution: for example, the French and German *Naturphilosophie* has embraced evolution since the early nineteenth century, and extended metaphysics to a geological scale. But their metaphysics precluded natural selection, which would do away with any need for the Platonic Idea, or archetype. To the Platonist, evolution must result from the progressive development of organic beings under the influence of the ideal form. It must be a kind of unfolding of that which has always existed in the timeless world of ultimate reality. It is for this reason that the very idea of evolution worked its way into biology by way of embryology, where a preexisting pattern is actually realized. To the *Naturphilosophie* of the last century, and to its present-day heirs as well, progressive change is not to be understood as a strictly natural process. Indeed, to call Goethe or Oken a forerunner of Darwin is like equating faith healers with practitioners of pyschosomatic medicine.

To an Aristotelian like Linnaeus, who believed that classes are real, there is no dichotomy between the order in nature and the system of classes expressing that order. The reality of classes implies that they are distinct and immutable, and excludes the possibility of evolution, so that a progressive change implies that there must have been successive creations. The goal of classification is reduced to discovering the underlying order, from which the system of classes follows automatically, for the two are basically identical. One discovers the classes, finds properties which separate one class from another, and erects a series of definitions.

Indeed, definition is looked upon as the ultimate goal of all classification, and a failure to depart from this tradition has long been a source of confusion in taxonomy.[8] Even now, systematists are inclined to believe that they "define" species rather than their names and to hold that if one cannot give a rigorous "definition" for the name of a group, then something must be wrong with the classification. Confounding definition of terms with "definition" in the sense of identifying or discriminating between entities, they fall into the error of trying to define the things instead of the symbols.

An understanding of how such metaphysical ideas have been an obstacle to the development of positive knowledge puts us in a position to consider Darwin's views on natural systems. Much light is cast upon Darwin's statements when we realize that he insisted on Aristotelian definition as a criterion of reality or naturalness. To Darwin, as to many other taxonomists, an inability to give rigorous definitions for the names of taxonomic groups led to a belief that somehow such assemblages were artificial. As a result, he came to stress the arbitrariness of the boundaries between species and to favor a genealogical standard for higher classification. In *The Origin of Species* he argues that our ability to group organisms into a hierarchy is a consequence of their genealogical relationship or "propinquity of descent." He therefore concludes that

> the natural system is founded on descent with modification; that the characters which naturalists consider as showing true affinity between any two or more species, are those which have been inherited from a common parent, and, in so far, all true classification is genealogical; that community of descent is the hidden bond which naturalists have been unconsciously seeking, and not some unknown plan of creation, or the enunciation of general propositions, and the mere putting together and separating objects more or less alike.[9]

Again, in summing up the expected impact of his theory, he argues that "Our classifications will come to be, as far as they can be so made, genealogies; and will then truly give what may be called the plan of creation." [10] Thus, Darwin clearly recognizes that evolution generates a "real" system of hierarchical relationships, and in a sense he equates "natural" with "genealogical." In so doing, he solved a very important problem, for prior to that time, "natural" was defined solely on the basis of metaphysical posits. This is true in spite of the fact that earlier writers had made a partial differentiation between "natural" and "artificial." Thus John Stuart Mill writes in his *Logic:* "The ends of scientific classification are best answered when the objects are formed into groups respecting which a greater number of general propositions can be made, and those propositions more important, than could be made respecting any other groups into which the same things could be distributed." [11]

These criteria provide us with a sensible, if vague, statement of the reasons why we seek natural systems. But the problem is swept under the rug by substituting one set of undefined terms for another. It treats the universe as if it were so constructed that a finite number of generalizations can be made about it. The very notion that entities have a limited number of commeasurable attributes, in the same sense that organisms are composed of so many cells or atoms, is a purely gratuitous assumption which has yet to be rooted out of taxonomy. And the issue of deciding what is important is left unexamined. Darwin solved the problem by redefining "natural" as derivative of the mechanism which underlies what was previously a mere empirical generalization about observed properties of organisms. The change he made exemplifies a basic shift in attitude. Instead of finding patterns in nature and deciding that because of their conspicuousness they seem important, we discover the underlying mechanisms that impose order on natural phenomena, whether we see that order or not, and then derive the structure of our classification systems from this understanding. The difference, then, lies with the decision as to what is important. It reflects the basic gulf in attitude separating idealists given to the older forms of induction, on the one hand, and empiricists who employ the hypothetico-deductive method, on the other. Classification ceased to be merely descriptive and became explanatory.

But the redefinition of "natural" created great difficulties, both practical and theoretical. One problem was with the two senses of the term "natural," for the redefinition affected both of these. Insofar as "natural system" means the order upon which the classification is founded, there is only one natural system, and its hierarchical order is the standard for the truth of any hierarchy of classes. But one must not neglect the other sense of "natural system," the general term for systems based on the genealogical nexus. In empirical classification as Darwin conceived of it, a number of natural systems in the second sense are possible. This ambiguity was never crucial for Aristotelian and Platonic systematics; it has posed great difficulties for modern biology. Although a system of genealogical relationships may be expressed in terms of the Linnaean hierarchy, the resulting classification is not as informative as one might wish.[12] Further, classification systems express more than genealogical relationships; they also have reference to degrees of change. Darwin took great pains to explain this point in *The Origin of Species*, where he writes:

> I believe that the *arrangement* of the groups within each class, in due subordination and relation to the other groups, must be strictly genealogical in order to be natural; but that the *amount* of difference in the several branches or groups, though allied in the same degree in blood to their common progenitor, may differ greatly, being due to the different degrees of modification which they have undergone; and this is expressed by the forms being ranked under different genera, families, sections, or orders.[13]

What this means is expediently clarified by an example from bird classification. Genealogically, birds are more closely related to crocodiles than to mammals, and for this reason it would be possible to include them as an order of reptiles. But they have undergone considerable modification, involving the rearrangement of numerous organ-systems and the assumption of an entirely new way of life. For this reason, they are given a higher "categorical rank" and made a class—equal in value to reptiles and mammals. In allowing such an alternative, the system is made more informative, yet it does not become artificial, for the modification of birds is an empirical fact. This is by no means the same as trying to base the system on an undefined degree of resemblance. Yet it is true that a certain amount of rigor has been lost in the use of the term "natural." In considering such widely different and scarcely comparable attributes as those characteristic of birds and mammals, there would seem to be no objective criterion for deciding what degree of modification justifies an increase in rank. When one group has changed in its reproductive system, another in its skelton, the best one can do is to have recourse to some arbitrary standard, since the two organ systems are not commeasurable. The basic structure of the Linnaean hierarchy is such that the genealogical order does not of itself provide sufficient criteria for ranking groups. Relationship and degree of modification are able to coexist in evolutionary systematics, but the stress laid upon them differs from one worker to another. Darwin, for example, chose to include the cirripedes among the Crustacea, in spite of the fact that they have assumed a drastically altered morphology and way of life. Owen, by contrast, stressed the differences and made the cirripedes equal in rank to the Crustacea.[14] One wonders if Darwin might have preferred to treat the birds as dumpy, toothless lizards, for one could say that he classified the barnacles in much the same way. Whatever standard one does take for ranking taxonomic groups, it should be clear that sytematists work at cross purposes when they do not agree on any such criteria. If a common standard were recognized, the system would be more informative by far, and the goal of natural classification would be better served.[15]

But leaving aside the problem of ranking, we may observe that Darwin had proposed a radical solution to the traditional question of the "reality" of taxonomic groups. What is "real" is the genealogical nexus, and the groups, or taxa, are chunks, so to speak, of this nexus. As a consequence, it became possible for the nominalist (and Darwin was something of a nominalist) to look upon taxa not as universals but as particulars, or individuals. Such a way of dealing with the problem raises philosophical problems, but we may overlook these for the moment. Under this point of view, the name of a group (taxon) designates a single, unanalyzed genealogical entity, rather than a class of similar things. Since the taxon is an individual, its metaphysical reality ceases to be a problem; and the adoption of a genealogical standard is largely motivated because of a desire to treat taxonomic groups as particulars. As Darwin put it in one of his notebooks: "Genus must be a *true cleft* putting out of case the analogys [sic]. —If genus does not mean this it means nothing." [16] Thus to Darwin, a taxon is real because it is a clade ("cleft") or genealogical unit—it is not a class name for a set of individual organisms with certain intrinsic properties in common. Uncertainties over this distinction have caused much philosophical confusion. The Linnaean hierarchy

has a form that permits one to treat the names of taxa as if they
were no more than intensionally defined class names—that is, as if
Vertebrata were simply a collective term for all backboned animals. But when the taxa are conceived of as founded by definition
on the genealogical nexus, the ability to treat them as intensionally defined classes may be purely accidental. The historical
processes of evolution generate a genealogical nexus which happens to be expressible in hierarchical forms of classification. On
the other hand, there are still purely nominal classes in the
Linnaean hierarchy—these are the *categories*. Consider the following series:

Superfamily	Hominoidea
Family	Hominidae
Genus	*Homo*
Species	*Homo sapiens*

The genus, the family, the superfamily, and so forth, are categories, the members of which are individual taxa, such as *Homo
sapiens*, Hominidae, and Hominoidea. *Homo sapiens*, a biological
species, is an individual in the class (category) of species, and
in the class of genealogical entities. It need not be looked
upon as simply a subgrouping in the "class" *Homo*, because
it is also an unanalyzable unit. John Smith is a part of *Homo
sapiens* in the same sense that John Smith's arm is a part of
John Smith.[17] There is a whole-part relationship, as well as one
of class inclusion. It is thus only a manner of speaking when we
treat a family as if it were no more than a class or set of genera.
Although the distinction between categories and the taxa which
are their members has helped to clear up much philosophical
confusion, there is still a tendency to look upon taxa as if they
were, like categories, strictly nominal. One cannot coherently
deny the reality of a species on the grounds that it is a class,
hence not "real," because there is a sense in which the group is an
individual. The same would be true of any entity with hierarchi-
cally structured relationships between its parts and for which a
set of classes was constructed in order to classify the components.
The United States of America is an individual in the class of
national states. To a nominalist at least, the class of national states
is not real, but the United States of America, being more than
just a class, is real. Similarly, the United States of America is

divisible into a number of states, such as California, and each of these in turn into counties. Now, "the national state," "the state" and "the county" are, like the categories of taxonomy, universals, there being no whole-part relation between them, and a consistent nominalist would have to deny their existence. But the United States of America, California, and Los Angeles County may be treated as particulars, and hence as real. It is therefore a mistake when certain radical nominalists have spoken of species and other taxa as "universals" having no real existence in nature; they would have to deny the existence of California as well. There are other ways, however, to solve the traditional problem of the "reality" of species, some of which are more philosophically conventional. Perhaps the best is to accept the "reality" of relational properties, which essentially amounts to the abandonment of nominalism. For the present analysis, it does not matter how we solve the problem; it matters only that the issues be understood. The important point is that a scientist's attitude toward classification systems is profoundly affected by the structure of his language, often in a manner of which he is quite unaware. That we tend to conceive of species as purely mental constructs is an accident of our culture. It is only a convention that the English language treats "society" as an abstraction and individual men as real. In certain other natural languages, society is "real," while the biological individuals are conceived of as abstractions.[18]

In making taxa something more than mental constructs, problems of identification became crucial. The name had to be related to the things which it designates, yet one does not define it, except derivatively, in terms of the characters which are useful in identification. The certainty which attaches to artificial classification systems is abandoned in favor of having the names refer to something of more fundamental importance. There are many who object to this maneuver and wish to continue in the tradition of naming taxa as mere classes of similar things. They wish to retain that feeling of certainty which accrues with our ability to classify organisms according to the presence or absence of particular intrinsic properties. But the ability to let the mind rest in arbitrary definitions is purchased at the price of triviality. Systematics has scientific value as explanation, not as mere description. The purpose of a classification is not the accurate pigeonholing or identification of enzymes or dried specimens, but the assertion of meaningful propositions about laws of nature and particular events. Those who abandon theoretical relevance for ease of definition and identification effectively desert from the ranks of science. Thus, the nineteenth-century philosopher Schopenhauer could hardly have understood biology's goals when he wrote: "Linnaeus adopted a vegetable system of an artificial and arbitrary character. It cannot be replaced by a natural one, no matter how reasonable the change might be, or how often it has been attempted to make it, because no other system could ever yield the same certainty and stability of definition." [19] But definition is of symbols, not of things or relationships, and symbols are useful only insofar as they are able to convey meaningful assertions about the subject matter of our discourse. To make the truth subservient to the whims of lexicographers is hardly worthy of a philosopher. The natural system eclipsed the artificial in scientific botany because it could serve as a fruitful instrument of scientific investigation. We can understand why Darwin wrote to Huxley: "Grant all races of man descended from one race—grant that all the structures of each race of man were perfectly known—grant that a perfect table of the descent of each race was perfectly known—grant all this, and then do you not think that most would prefer as the best classification a genealogical one, even if it did occasionally put one race not quite so near to another, as it would have stood, if collocated by structure alone?" [20] Here Darwin argues that theoretical relevance is not to be traded for convenience of identification. It would appear that Huxley had objected that genealogy is not knowable with certainty, a view which is still occasionally voiced by radical empiricists. But such criticism merely restates the well-known truth that inductive inferences are never conclusively verified. Another, equally unfounded objection to phylogenetic systematics is that structural attributes are used in identification. It is thus asserted that because organisms are identified on the basis of characters—morphological or otherwise—there is some reason for considering taxa to be artificial classes, defined intensionally in terms of the characters useful in identification. But this view may be refuted easily, by considering how we do, in fact, use one set of criteria for definition of names and another for identification of the things which the names designate. Thus, the word "father" means some man whose sperm has united with an egg to give rise to a child, but our assertion that someone is in fact a father is made because of indirect evidence. If we say that someone is a father, we are right or wrong solely if it is materially true that he has engendered offspring, and irrespective of whether or not evidence is availble.[21]

DARWIN AND THE SPECIES PROBLEM

It has already been demonstrated that Darwin considered "propinquity of descent" the basis of natural classification. His motives have been explained as the consequence of a particular kind of nominalism: a taxon is a genealogical entity, not a class of morphologically similar organisms. For this reason it seems inconsistent that Darwin should be ranked among those who deny the "reality" of species. Nonetheless, it is generally agreed, even by authorities whose judgment deserves the highest respect, that Darwin advocated a morphological species concept—that he believed that species should represent degrees of similarity.[22] And those who wish to deny that species are real often appeal to Darwin for support.[23] Controversies as to the meaning of the term "species" have spawned an enormous literature.[24] The "species problem" is a curious mixture of semantic confusion, significant questions as to the utility of one or another verbal convention, and logical mistakes of considerable variety. One's stand on such matters is in part determined by his philosophical outlook. To Linnaeus, there was no species problem, because Aristotelian philosophy, as he understood it, implies that the universe is organized in the form of classes. To nominalists, it was possible to look upon species as corresponding each to a group descended from progenitors created by God in the beginning. Or, the very existence of species could be denied altogether, and the evident distinctness of organisms in nature could be viewed as illusory. With the discovery of evolution, involving the historical origin of one species from another, the existence of a distinct gulf between all species at all times could not be maintained. A shift in the scientist's conception of species became mandatory.

In his efforts to grapple with the implications of his theory, Darwin lacked many of the most significant insights that have been provided by subsequent research. In modern biology, partly because of investigations into the properties held in common by what had been called species, and partly owing to our understanding of the mechanisms of evolution, the "biological species concept" has increasingly prevailed. We now know that pre-evolutionary taxonomy was discovering a real order when it grouped organisms into species. A species is a population, a unit of evolution and of reproductive activity—a kind of social entity. It comprises all those biological individuals which exchange genetic material with one another, and which are reproductively isolated from organisms in other populations of the same nature. It is thus an integrated unit of biological function, rather than a mere class of similar things.

But to so define "species" requires a high level of abstraction—one which is often overlooked, with grave consequences. The biological species concept entails the use of an idealized model of the underlying processes, a model which is perhaps best compared with the "perfect gas" of the perfect-gas laws.[25] If one imagines a population of organisms becoming separated by a geographical barrier into two distinct subunits, and if one further imagines the two derived populations undergoing different patterns of evolutionary modification, one can see how, by changes in reproductive physiology or behavior, it would become increasingly difficult for the individuals in one group to pair with those in the other and produce viable offspring. Early in the process, the dissolution of the barrier would result in an exchange of genetic material, and any difference that had built up between the component organisms of the two populations would disappear. But the differences developed in isolation should gradually become so great that the derived populations would be prevented from interbreeding; or, more technically speaking, they would become reproductively isolated. The process of isolation and the development of a barrier to reproduction is called *speciation*, and the products resulting from it are, by definition, *biological species*. Speciation is the reason why it was possible to delimit what were already called species before the term "species" was thus redefined—it is the causal basis for resemblances, constancy, distinctness, and other criteria by means of which their existence was inferred. Thus, modern biology does not use the term "species" to designate a different kind of entity; it has only substituted more rigorous defining properties of a scientific explanation for the mere empirical recognition of discontinuities.

The use of an idealized model as a standard of definition creates serious difficulties in practice. The actual course of events is rarely so simple as to permit drawing a distinct line between one species and another, especially since the status of a population as a species is a matter of degree. There are a host of practical problems whenever one attempts to decide if reproductive isolation has in fact arisen. Nor does the biological species definition tell one what to do with forms which no longer reproduce sexually, or with a single population that becomes altered in a geological series. But such difficulties in no way invalidate the definition, for many words give much the same problem. For instance, the concept of a biological individual breaks down when we try to distinguish between parent and progeny or to delimit individuals in a colony of hydroids. The decision to organize our thought in terms of biological individuals and species derives, not from our ease in distinguishing between such entities, but from our knowing that they do in fact exist and from our realizing the expediency of having a language which allows us to discuss them. Therefore the definition of terms in this manner is conventional, but it in no way follows that one definition is as expedient as another for the ends we have in mind. And once one sees the distinction between the definition of abstract class names and the application of such concepts in practice, the species problem becomes a pragmatic one.

It is not difficult to find passages in Darwin's writings suggesting that he considered species to be purely artificial constructs. For instance: "In short, we shall have to treat species in the same manner as those naturalists treat genera, who admit that genera are merely artificial combinations made for convenience. This may not be a cheering prospect; but we shall at least be freed from the vain search for the undiscovered and undiscoverable essence of the term species." [26] On the basis of statements such as this, Mayr and others have come to the conclusion that Darwin upheld a morphological, as opposed to a biological, species concept.[27] That is, they maintain that he looked upon species as merely classes of organisms having a given degree of similarity and difference in the observed properties of their members.

But there are certain other passages in Darwin's writings which show that the problem is more complicated. In his second notebook on the transmutation of species, Darwin says: "As *species* is

real thing with regard to contemporaries—fertility must settle
it." [28] The reality of species is affirmed in a letter to Gray written in 1860, in which Darwin severely criticizes some assertions
of Louis Agassiz. The following statement is particularly relevant
to the question at issue: "How absurd that logical quibble—'if
species do not exist, how can they vary?' As if any one doubted
their temporary existence." [29] It appears that there is at least one
sense in which species are thought to be real, although it is evident that there is a sense in which they are held to be not real.
This being the case, any citation of statements by Darwin in
support of his holding one or another point of view must be
buttressed by a demonstration of the sense which he intended.

That Darwin did not look upon species as necessarily wholly
arbitrary and founded merely upon morphological distinctness
and the like is demonstrable from his actual procedure in systematic work. In a letter to J. D. Hooker, dated 30 March 1859,
Darwin says: "As for our belief in the origin of species making
any difference in descriptive work, I am sure it is incorrect, for I
did all my barnacle work under this point of view. Only I often
groaned that I was not allowed simply to decide whether a difference was sufficient to deserve a name." [30]

It is thus clear that Darwin upheld the reality, in a sense, of
species both in theory and in practice. In order to see how he
thought of them as real, it will be convenient to first eliminate the
senses in which he held that they are not real. One such conception was taken over, quite directly it would seem, from Lyell,
who argues that for a species to be real, it *must* be constant in
character.[31] This would follow quite consistently from the
Aristotelian tradition, which implies that names can only be defined disjunctively—that is, in terms of some distinct and invariant set of differences between every member of each class.[32] If
there has been evolution, the species must have graded into each
other, and a distinct gap would not be available for drawing a line
between the ancestral and the derived species. Hence, the two
species could not be "defined" and therefore they do not exist in
nature. This attitude is clearly expressed when Darwin says:
"Every naturalist who has had the misfortune to undertake the
description of a group of highly varying organisms, has encountered cases (I speak after experience) precisely like that of man;
and if of a cautious disposition, he will end by uniting all the

forms which graduate into each other, under a single species; for
he will say to himself that he has no right to give names to objects
which he cannot define." [33]

There is yet another sense in which Darwin held that species
are not real. He maintained that there are no "essential" differences between species and varieties, and that both terms designate
the same basic kind of entity. Thus Darwin says: "I look at the
term species, as one arbitrarily given for the sake of convenience
to a set of individuals closely resembling each other, and that it
does not essentially differ from the term variety, which is given
to less distinct and more fluctuating forms." [34] This statement
means, not that the boundary between one species and another is
arbitrary, but rather that there has been no convention for determining whether an entity is to be ranked as a species or a variety.
This must be Darwin's intent when, after discussing the question
of how many species of human beings there are, he remarks:

> But it is a hopeless endeavour to decide this point, until
> some definition of the term "species" is generally ac-
> cepted; and the definition must not include an indetermi-
> nate element such as an act of creation. We might as well
> attempt without any definition to decide whether a cer-
> tain number of houses should be called a village, town,
> or city. [35]

It is a perfectly valid point of view which Darwin expresses, in
asserting that there are certain arbitrary elements in classification.
But such objections in no way conflict with the idea that the taxa
are real. There is no statement in the entirety of Darwin's published writings which, properly interpreted, asserts that there is
nothing more than a name between the individuals of a species.
For, in technical terms, Darwin was denying the reality, not of
taxa, but of categories. We need only refer, in confirmation, to
the last quotation, which asserts the arbitrariness of the line
which we draw between the town and the city, but says nothing
about Oxford and London. It is not at all the same thing to deny
the existence of "the species" on the one hand, and of *Homo
sapiens* on the other. The existence of populations with the attributes necessary and sufficient to make them species by definition is an empirical fact.

Part of the reason why Darwin's views on the species have so

often been misinterpreted, is that he bent over backward to em-
phasize both borderline cases and the mutability of species. For
purposes of dialectic, he laid great stress on our inability to draw
a clear line at the stage at which a variety becomes a species. The
special-creationist viewpoint implied that species are very much
distinct from one another. To the special-creationist, "the term
includes the unknown element of a distinct creation." [36] If each
species had been created as a single pair, one would expect provi-
sion to have been made for preventing sexual union with the
wrong partner. The existence of physiological sterility barriers
between distinct species would therefore be expected by special-
creationists; and the criterion of sterility was in fact invoked by
such special-creationists as John Ray (1686).[37] But there is a
profound gulf between the sterility test of pre-Darwinian biologists and the reproductive isolation of modern biology. If species
were specially created, one would expect the inability to cross to
be as absolute as the gap between the species themselves. But the
hypothesis of natural selection implies that there should be all
degrees of sterility as a direct consequence of the gradual development of all properties. It was therefore very important for
Darwin to demonstrate that the ability to cross displays no invariant correlation with the degrees of morphological similarity
according to which systematists had traditionally grouped animals and plants.[38] For this reason alone, we should appreciate the
pains Darwin took to show how sterility barriers break down
under domestication, how artificial crosses may be produced
where none occur in nature, and how the sex of each of the
crossed forms affects the degree of sterility. And the unreliability of the sterility test, as well as its lack of strict correlation with
morphological divergences, is implicit in the theory of natural
selection. The properties which develop during speciation are the
result of a number of independent processes. It is not strictly true
that one many take a sample of individual characteristics which
distinguish one population from another and on this basis say with
certainty that speciation has occurred. For in those instances
where selection pressure acts chiefly upon the reproductive system, one may anticipate the rapid development of reproductive
isolation, with "cryptic species." Furthermore, reproductive isolation may develop by a variety of mechanisms—morphological,
immunological, behavioral—none of which is strictly comparable

to the others. The redefinition of the word "species" therefore, so as to make it the basic unit of evolution, should in general lead to the formulation of groups displaying some correspondence to morphological "species," but this need not be the case.[39] In criticizing the sterility test, Darwin was not asserting that it should not be used; he was only saying that a morphological criterion is not always consistent with that of sterility. He did this, not for the purpose of defining "species" as a term, but to deny their supposed immutability. Together with the distinctions already made as to Darwin's views on the reality of species, the fact that he was criticizing inconsistent usages suffices to clear up many of the misconceptions of his opinions. Thus, the following statements from the second chapter of *The Origin of Species* do not propose how terms should be employed, but merely describe common usage—they are strictly lexical definitions:

> Hence, in determining whether a form should be ranked as a species or a variety, the opinion of naturalists having sound judgement and wide experience seems the only guide to follow. We must, however, in many cases, decide by a majority of naturalists, for few well-marked and well-known varieties can be named which have not been ranked as species by at least some competent judges.[40] . . . I look at the term species, as one arbitrarily given for the sake of convenience.[41] . . . but the amount of difference considered necessary to give to two forms the rank of species is quite indefinite.[42]

Darwin's rejection of the sterility test is in many ways a consequence of his having abandoned essentialism. It also bears some analogies to his attitude toward the *scala naturae*, with its "higher" and "lower" organisms. Evolutionary diversification results from processes that affect different kinds of structures, with the various parts evolving at different and changing rates. Only when one presupposes that the various kinds of evolutionary events are commensurable is it possible to establish any number of intrinsic properties of biological individuals as determining specific, or other, categorical rank. Such equivalence exists for some Platonists and Aristotelians, but makes no sense without the typological premises on which these metaphysical systems rest. It is reasonably sound induction to use a comparison of particular morphological properties as evidence that a certain amount of change has occurred. But the idea that there can be a meaningful morphological measure of overall evolutionary change or "taxonomic distance" is a delusion inherited from an invalidated manner of thinking, for changes do not take place "overall." It is partly for the reason that it has no such drawbacks that the biological species concept has prevailed. The delusive quest for a set of intrinsic properties of biological individuals as a criterion of specific rank has been abandoned in favor of attributes of the population. The boundary between varietal and specific status is now seen to be distinct and due to a single cause. The differentiation is no longer formal or morphological, but is grounded in function. It employs a dichotomy which, like that between foetus and infant, is conventional, but which nonetheless has profound biological significance.

We may now turn to an analysis of Darwin's views on speciation and see how they compare with modern concepts. His first notebook on the transmutation of species suggests some distinct parallels: "A species as soon as formed by separation or change in part of country, repugnance to intermarriage—settles it." [43] Analogous statements from his later writings demonstrate that he never departed from this point of view.[44] Modern evolutionary theory envisions speciation as occurring by a process of isolation followed by such changes as would prevent subsequent interbreeding. The main difference between this early (1837 or 1838) view given by Darwin and the modern one is Darwin's insufficient awareness of the importance of geographical separation.[45] However, this is largely a matter of emphasis, for Darwin was quite aware of the fact that spatial isolation provides conditions conducive to species formation. This is clear from a letter of 1862 to J. D. Hooker, in which Darwin argues: "If 1,000 pigeons were bred together in a cage for 10,000 years their number not being allowed to increase by chance killing, then from mutual intercrossing no varieties would arise; but, if each pigeon were a self-fertilising hermaphrodite, a multitude of varieties would arise. This, I believe, is the common effect of crossing, viz., the obliteration of incipient varieties." [46] This quotation shows how Darwin treated the problem of varieties in groups which, not being united by sexual reproduction, are by definition not reproductive populations—a problem which is still a difficulty for advocates of the biological species definition. It also seems clear from this statement that although Darwin was able to conceive of the process of speciation, he did in fact use the terms "variety" and "species" in the strictly morphological or descriptive sense. Yet it cannot be stressed too strongly that his traditional use of words does not in the least contradict his advocating, where feasible, a terminology founded on evolutionary processes and relationships. It would seem that he used conventional vocabulary and at the same time conceived of evolution and taxonomic groups in a more modern fashion.

Darwin's actual suggestions for the improvement of taxonomy are given in a section of *The Origin of Species* quite separate from the lexical accounts of usage which are generally quoted in support of his alleged morphological species concept.[47] The discussion follows, in both literary and conceptual sequence, his argument that systems of classification ought to be genealogical. He states that varieties are "incipient species" and that species, tending to diverge, form genera.[48] And he relates the hierarchical structure of taxonomic entities of lower rank to the same genealogical nexus, when he says: "The origin of the existence of groups subordinate to groups, is the same with varieties as with species, namely, closeness of descent with various degrees of modification."[49] Finally, he explicitly affirms the desirability, in his way of thinking, of an evolutionary classification at the lower levels and argues much as he does with respect to supraspecific categories: "In classing varieties, I apprehend if we had a real pedigree, a genealogical classification would be universally preferred; and it has been attempted by some authors."[50]

The argument for an evolutionary system of classification is immediately followed by a refutation of the purely morphological species concept, on the grounds that, even in practical systematics, biological unity invariably takes priority over morphological similarities and differences. Darwin begins with sexual dimorphism:

> With species in a state of nature, every naturalist has in fact brought descent into his classification; for he includes in his lowest grade, or that of a species, the two sexes; and how enormously these sometimes differ in the most important characters, is known to every naturalist:

scarcely a single fact can be predicated in common of the
males and hermaphrodites of certain cirripedes, when
adult, and yet no one dreams of separating them.[51]

Then Darwin presses his attack, by enumerating other polymor-
phisms such as alternation of generations, larvae *vs.* adults, and
the different forms of flowers that exist in a number of plant
species. The argument is carefully designed to demonstrate that
the morphological species concept conflicts with universal practice, leads to absurdities if carried to its logical conclusion, and
makes no sense as an expedient basis for the construction of a
biological language.

Some of Darwin's views on classification at the species level are
summed up in the final chapter of *The Origin of Species.* But
here the statements occur in the context of prognostications for
the impact of his theory on the future development of the sciences in general. He stresses, on the one hand, the artificiality of
taxonomic groupings—quite understandably, for once the genealogical and evolutionary standards had been accepted, metaphysical questions about the "essence of a species" would no longer be
relevant. Yet on the other hand he affirms that, in future practice,
the importance of differences should receive more attention. He
accepts the need to infer specific distinctness on the basis of a
lack of intermediate forms. But, he continues: "Hence, without
quite rejecting the consideration of the present existence of intermediate gradations between any two forms, we shall be led to
weigh more carefully and to value higher the actual amount of
difference between them." [52] This is a crucial point, for it makes
a basic epistemological distinction. In looking at a collection of
organisms, we can group the individuals together according to
the number of observed resemblances in common. When organisms evolve, the species do, in fact, tend to have certain properties that distinguish them from other species, in correlation with
the real evolutionary relationships. But how do we know, in
founding groups simply on those properties which strike the eye,
whether or not we have selected properties actually correlating
with the underlying order? Some would assert that there is no
way to make such a distinction and that classification must necessarily be based upon "overall similarity," "resemblance,"

"morphological distance," and the like. Thus, according to this viewpoint all classification is by necessity based on a psychological impression, rather than on objectively meaningful relationships. But to Darwin, there is a real order in nature, it is knowable in spite of the incertitudes of induction, and it can serve as a basis for classification, even at the species level. His reference to "the actual amount of difference" points up this distinction. A classification is natural not because of the available evidence, but because of the empirical truth of the propositions which it entails.

That the foregoing interpretations of Darwin's opinions are correct can thus be supported by their coherent explanation of his theoretical pronouncements. But they may also be verified by the manner in which he used such concepts in practice. We may take as an example the *Primula* question. The issue of whether these plants should be ranked as species or as varieties is a recurrent theme in Darwin's thought; indeed, it may be traced back to Lyell's *Principles of Geology*.[53] In *The Origin of Species*, *Primula* is mentioned in two different contexts. The first reference is in relation to the different usages of the term "species" and the difficulty in drawing a distinct line between species and varieties. Darwin says:

> Many of the cases of strongly-marked varieties or doubtful species well deserve consideration; for several interesting lines of argument, from geographical distribution, analogical variation, hybridism, &c., have been brought to bear on the attempt to determine their rank. I will here give only a single instance, the well-known one of the primrose and cowslip, or Primula veris and elatior. These plants differ considerably in appearance; they have a different flavour and emit a different odour; they flower at slightly different periods; they grow in somewhat different stations; they ascend mountains to different heights; they have different geographical ranges; and lastly, according to very numerous experiments made during several years by that most careful observer Gärtner, they can be crossed only with much difficulty. We could hardly wish for better evidence of the two forms being specifically distinct. On the other hand, they are united by many intermediate links, and it is very doubtful whether these links are hybrids; and there is, as it seems

to me, an overwhelming amount of experimental evidence, showing that they descend from common parents,
and consequently must be ranked as varieties.[54]

Later in the same work, immediately after his discussion of the
need to base species discriminations on the "actual amount of
difference," Darwin says: "It is quite possible that forms now
generally acknowledged to be merely varieties may hereafter
be thought worthy of specific names, as with the primrose and
cowslip; and in this case scientific and common language will
come into accordance." [55] This prediction was fulfilled by Darwin himself, who, in conjunction with his studies on the reproductive physiology of this genus, carried out a considerable number
of crossing experiments and gathered a massive body of additional data.[56] In 1869, he published a paper now quite generally
overlooked, entitled "On the Specific Differences between *Primula veris*, . . . and on the Hybrid Nature of the Common
Oxlip." [57] In this work he reviews the evidence for specific distinctness, stressing such points as the fact that the various forms
are adapted to fertilization by different insects. He enumerates a
number of crossing experiments, showing low fertility. He
demonstrates that the intermediate form, or oxlip, is a sterile
hybrid, and supports this influence by showing that the oxlip
occurs where the parent species are present, but not otherwise.
The third species is shown to be sterile when crossed with the
others, and to be distinct in morphology and in geographical
range. In summing up his observations, Darwin says that the
various forms "are all descended, from the same primordial form,
yet, from the facts which have been given, we may conclude that
they are as fixed in character as the very many other forms which
are universally ranked as species. Consequently they have as good
a right to receive distinct specific names as have, for instance, the
ass, quagga, and zebra." [58] When we realize that the analogy
with horses is an oblique reference to mules, we see that Darwin
was stressing reproductive distinctness. In other words, he did
recognize the desirability of bringing nomenclature into line with
biologically significant discontinuities, and he did make a deliberate effort to effect the reforms implied by the new standard. This
being the case, it seems reasonable to conclude that his actual
practice bears out the distinctions already made. Recognizing
what his views really were will not resolve any of the issues in

the species controversy, since these should be judged on the basis of intrinsic merit; but the time has come for an end to the invocation of Darwin's authority in support of a morphological species definition.

Let it not be thought, however, that Darwin supported the biological species definition in its strictly modern sense. There is no solid evidence that he conceived of species as reproductively isolated populations. His emphasis lay more with the distinctness of the individuals in different species in terms of their biologically important characteristics, and also with the genealogical interrelationships of the individuals within each species. Such a conception of species is somewhat more divergent from the biological one than might be thought. The biological species concept stresses the integrity of species as units of function. It emphasizes reproductive isolation because once a barrier to crossing has arisen, new units of interaction are formed, and for this reason the change from a geographic race to a species involves a fundamental alteration in the properties of the unit itself. The change is not absolutely discontinuous, but it has profound ecological and evolutionary significance; as with distinct stages in development of biological individuals, the differences are momentous. Those who object to this differentiation should, to be consistent, affirm that there is no real difference between fertilized and unfertilized eggs, between a foetus and a newborn infant, or between a living animal and a corpse.[59] Darwin seems not to have fully appreciated the importance of the discontinuity, and in this sense he did not embrace what we would call a biological species concept. However, he did recognize that there are species, and he did conceive of them as units or stages in the evolutionary process. Perhaps the evolutionary species concept of Simpson is the closest modern parallel to Darwin's: *"An evolutionary species is a lineage (an ancestral-dependent sequence of populations) evolving separately from others and with its own unitary evolutionary role and tendencies."* [60] Simpson argues that there is no real difference between the "biological" and "evolutionary" concepts, except that he considers the evolutionary one more comprehensive. To be sure, both reflect a desire to so define the word "species" that it will refer to evolutionary units having real existence in nature. But the biological species definition has the advantage of making a clear distinction between species and varieties. In addition, it provides a better recognition of the fact that populations are units of interaction. Still, unlike the morphological species definition, the evolutionary definition is by no means founded on an inappropriate conception of the goals of scientific research. It simply fails to take account of a very important natural discontinuity. For this reason, Darwin's views on the species must be looked upon as closely akin to those of modern biology. Accusing him of a typological attitude toward them is in no way justified.

The foregoing analysis has great relevance to our conception of Darwin's intellectual competence. For in spite of the fact that Darwin has repeatedly been dismissed as having little ability for abstract and speculative thought, it appears that a coherent interpretation may be made of his ideas on the philosophy of classification. There is every reason to think that his views on such matters are consistent throughout his works, and there is no real evidence of fundamental contradictions. Further, it is obvious that he dealt with some very difficult problems, both biological and philosophical. Because he was aware of the intricate logical issues with which he dealt, he was able to avoid many of the perennial mistakes which have plagued philosophers and biologists alike. Here, as always, Darwin's references to philosophical matters tend to be implicit and oblique. Yet he deals, and deals successfully, with a host of philosophical ideas. Indeed, one might conjecture that his philosophical prowess was in no small measure responsible for his scientific triumphs.

5. Barnacles

One of the unexplained mysteries of Darwin's life is the fact that he took up the study of barnacles. It seems curious that Darwin devoted some eight years to revising the classification of a highly modified group of crustaceans, when he had already developed his evolutionary hypothesis and had even written two preliminary drafts of a work expounding upon it. The mystery deepens when one reflects on Darwin's ill health: death could easily have precluded his publishing at all. The reasons given for this apparent deviation from his fundamental interest are usually one version or another of some platitudes originated by T. H. Huxley.[1] There is the notion that he needed to work with the organisms themselves in addition to speculating; this overlooks the greater utility of his experiments and observations on domesticated plants and animals. Or a need for self-discipline might be suggested—as if his earlier geological investigations had not provided enough. There is talk about the importance of practical work in discriminating species—which could have been obtained through a much smaller project. Finally, there is an alleged guilt over his being too much of a compiler—the absurdity of which is manifest from the fact that his geological writings, by no means compilations, were the major portion of his contribution at the time. To be sure, the experience which he obtained, even in the more tedious aspects of the work, was of great value, and he acknowledged that it helped him when he came to write *The Origin of Species*. Also, the degree to which organisms vary in nature was brought to his attention with particular forcefulness as he tried to sort out his materials into distinct groups. But any amount of experience is likely to have some utility, and Darwin hardly needed to spend eight years revising a whole sub-class. In this sense we may agree with Darwin in doubting "whether the work was worth so much time."[2] Yet in another sense one must object to such a harsh criticism, for his monograph has long remained one of the standards of excellence in taxonomy. Its interest is more than historical, for it is still an indispensable reference, and it continues to provide useful ideas for those who are carrying on the study of barnacles. The significance of the work as an innovation in methodology has scarcely been realized. To be sure, Darwin is recognized by botanists as a comparative anatomist of the first rank.[3] But Darwin's plant anatomy, if brilliant, is less original in technique than is his work on barnacles. Indeed, the latter is so novel that nobody seems to have understood it, for his method was as much a departure from that of his predecessors as was that of Vesalius in his *De Humani Corporis Fabrica*, or Harvey in his *De Motu Cordis*. Yet even this does not justify its taking priority over *The Origin of Species*.

It has been suggested that Darwin took up the study of barnacles because of their central position in a classification system elaborated by E. S. MacLeay.[4] This system was, in effect, a numerological one and therefore incompatible with the theory of evolution. It was based on the notion that all taxonomic groups may be arranged in sets of five subgroups, arranged in a circle. One might think that overthrowing this system would remove one objection to Darwin's theory. But the notion that Darwin had to refute it before his own hypothesis would be acceptable can scarcely be argued coherently. It is true that Darwin mentions MacLeay's system in his notebooks on transmutation of species.[5] But a working biologist would hardly conceive of a particular classification system, and a most heterodox one at that, as a serious objection to a comprehensive theory consistent with essentially all other classifications in use at the time. To refute the "Quinary" system, as it was called, was no more necessary than answering Paley's natural theology point by point. And although Darwin ultimately did go on to refute a number of marginal thinkers (including Paley), he did so only after *The Origin of Species* had been published.

It appears that Darwin's original intent with respect to the barnacles was only to describe a single species, as a small, personal contribution to the body of taxonomic literature based on collections from the *Beagle* voyage, much as he had done for a few flatworms that had interested him.[6] That Darwin did not intend to lavish so many years on the project is clear from a letter to Hooker, of 1845, in which he says: "I hope this next summer to finish my South American Geology, then to get out a little Zoology, and hurrah for my species work. . . ." [7] The particular barnacle which drew his attention to the group was morphologically unusual, as he explains in his autobiography:

> In October, 1846, I began to work on "Cirripedia." When on the coast of Chile, I found a most curious form, which burrowed into the shells of Concholepas, and which differed so much from all other Cirripedes that I had to form a new sub-order for its sole reception. Lately an allied burrowing genus has been found on the shores of Portugal. To understand the structure of my new Cirripede I had to examine and dissect many of the common forms; and this gradually led me on to take up the whole group.[8]

Darwin asserts that the valuable and interesting aspect of the work was not the description of species, but the anatomy.[9] It thus seems reasonable to infer that Darwin was interested in certain problems of comparative anatomy, perhaps in relation to evolution. If so, his revision of the group is completely intelligible, as it would have been necessary for the corroboration of his anatomical insights. In addition, an ability to use evolutionary theory as a guide to the discovery of taxonomic relationships would be one way of testing his hypotheses and of dispelling any doubts that might have remained. There are some difficulties, by no means insoluble, when one attempts to support such conjectures about Darwin's taxonomy. Nowhere in his writings on cirripedes does Darwin discuss, in explicit terms, the relationships between evolutionary theory and the biology or systematics of barnacles. His references to such matters are there, but covert, scattered, and cast in the form of an older terminology, for these works were published before *The Origin of Species*. Nonetheless, one may, by careful reading, and especially by analogy with statements occurring elsewhere, reconstruct his ideas on comparative anatomical method and the evolution of barnacles and demonstrate their connections with the rest of his thought.

COMPARATIVE ANATOMY BEFORE DARWIN

When Darwin began his researches on barnacles, comparative anatomy was already a well-developed science. Indeed, its fundamental operation—point-by-point comparison of similar parts throughout a variety of organisms—may readily be traced back to the writings of Aristotle. As early as 1555, Belon published a pair of drawings showing the skeletons of a man and a bird with the homologous, or morphologically equivalent, bones labeled with the same letters.[10] Subsequently one may detect the continuing development of a major theme in comparative anatomy: the attempt to treat diverse organic beings as if they were constructed according to a single plan. The trend reached its culmination around the turn of the nineteenth century, with the effort to found an independent science treating the strictly formal properties of organisms. For this branch of learning the poet Goethe (1749–1832) coined the term "morphology," the goal of which was to abstract the underlying pattern, or archetype, as it came to be called. The study of pure morphology is still very much alive, and many of the methods worked out in pre-Darwinian times remain in use. Morphology reached a peak of theoretical interest with the attempt of Etienne Geoffroy Saint-Hilaire (1772–1844) to demonstrate that there is a single plan of organization for all animals, and with Goethe's argument that all the organs of plants are essentially variant forms of leaves. Since then, pure morphology has declined, in the sense that it has become fused with other disciplines.

One of the reasons for the decline of pure morphology was its failure to stand up to valid criticisms. Cuvier (1769–1832), one of the founders of modern comparative anatomy and of paleontology, in a very celebrated debate with Geoffroy, was able to take the assumption that there is a single plan of organization and to show that it led to contradictory conclusions.[11] This *reductio ad absurdum* implies that the abstraction of a scheme for comparison is not to be conceived of as an end in itself; it is an aid to the solution of more basic problems, useful only insofar as the many exceptions do not obscure the truth. And our ability to formulate it tells us nothing until it is linked up with an explanation. Cuvier had yet another criticism of purely formal morphology, which was that the characteristics of organisms are in close correlation with habitat and with the "final causes" of the parts (or, as he put it, with the "conditions of existence"). To Cuvier, the ideal system of classification was not morphological but physiological. And many would agree that a classification which relates the properties of organisms to a system of physiological explanations is more meaningful and informative than an abstract scheme of formal relationships. The shift in emphasis is typical of the contrast between ancient and modern science: to Aristotle, the final and formal causes were most important; now we value the material and efficient. Yet the worst defect of the old morphology was that it concerned itself with a search for something which does not exist. Archetypes are Platonic Ideas; they are strictly metaphysical constructs. Goethe's enthusiasm for morphology was typical of his attitude toward all branches of human experience. He believed in a world of superior reality: a *beau ideal* in aesthetics, a plant morphology organized in terms of an ideal plant or leaf, and an established order in society.[12] His idealism even led him to reject Newton's theory of color, and he tried to replace it with one which could interconnect our subjective experiences of apprehended colors with the ideal world, much as our apprehension of day-to-day experience is idealized in sculpture or poetry. The abstractions of Newton were rejected because they do not mirror our subjective experience with sufficient vividness, and are far too removed from our inner life. Thus the errors of Goethe's *Farbenlehre* grew from the same metaphysical roots as did his ideal plants. In the theory of color, it is obvious that Goethe was wrong—but his false premises have yet to be purged in their entirety from morphological thought.

It is easy to find elements of the older comparative anatomy in Darwin's work on barnacles. Indeed, the traditional methods based exclusively on formal properties have not been wholly abandoned to this day. The difference lies in shifting from enumerative induction to a more hypothetico-deductive approach and in recognizing new kinds of argument. The use of formal properties has already been discussed in Chapter 1, where it was shown how one may use the intrinsic properties of rocks to compare sequential relations in groups of strata. Such comparison has obvious bearing on questions of historical process, and can be applied in other situations, including anatomy. Darwin's

monograph includes a set of diagrams (here reproduced as fig. 4),
showing the arrangement of parts in the plates which cover the
sides of a typical "acorn" barnacle.[13] The diagrams, adapted
from the work of one of his predecessors, are a traditional repre-
sentation of what he called a "homological plan," depicting only
certain features common to the known members of a group and
not necessarily asserting anything about historical origins. Each
basic part in any form can be seen to correspond to a similar part

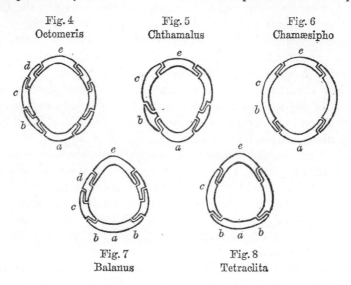

a, Rostrum; *b,* Rostro-lateral, *c,* Lateral, *d,* Carino-lateral compartment; *e,* Carina

FIG. 4 The arrangement of external plates. Note the different pattern of
fusion in Figs. 4 to 6 (Chthamalinae) and 7 and 8 (Balaninae). From
Darwin's barnacle monograph.

(such as a whole plate or an articulation) in some other species,
although some parts may be absent or fused. The corresponding
parts (technically, "homologues") are labeled with the same letter. What the depicted correspondence means has long been a
source of confusion in the philosophy of biology. To a Platonist,
the plates can all be related to an ideal organism, and the relation
between them is one of "essential similarity." To Darwin, and to
the modern evolutionary systematist, such an interpretation is
nonsense. There is no essence, and therefore nothing can stand to

it in the relation of essential similarity. To a radical nominalist, to say that two plates are homologous simply means that they are similar. But similar in what? The relation of similarity must have its terms supplied or it means literally nothing. In Darwin's figures, plates *a* and *e* are similar in terms of some properties, yet different in others. To say that they are "just plain similar" is as ridiculous as saying "John is taller than." Further, there are certain other resemblances that need to be distinguished. It is clear, for example, that there are basic differences between plate *c* in Darwin's figure 6, and that designated by *c* and *b* in his figure 8. But, owing in part to their having much the same function, they share certain attributes. Thus one must distinguish between a more basic kind of correspondence and that which accompanies a superficial resemblance—between what are technically called "homology" and "analogy." The mere notion of similarity is thus as sterile as is an undefined essence.

THE DARWINIAN REVOLUTION IN COMPARATIVE ANATOMY

Darwin brought about a new way of thinking about morphological comparisons: he treated the archetype as a consequence of past evolutionary processes. He deliberately ridiculed the conservative anatomists who continued to look upon such schemes as ideas in the mind of God. In consequence he effected a revolution in the attitudes of many systematists toward their work. The construction of archetypes gave way to the reconstruction of common ancestors, and systematics became inseparable from "phylogeny," a historical science of organic development. Yet there are those who have argued that the difference has been exaggerated.[14] It is contended that a reconstructed common ancestor is nothing more than an archetype with a new name, and that a phylogenetic sequence is just another way of expressing a taxonomic hierarchy. In other words, it is held that evolution only explains typological systematics while adding nothing to its method. Such views, though often repeated, have never been supported by empirical arguments. Indeed, the notion that the Darwinian revolution had no effect on the work of taxonomists is a universal negative and therefore cannot be verified. Such notions, which derive from essentialist metaphysics and from a failure to understand the modern scientific method, can easily be refuted by reference to the facts; and the works of Darwin provide almost ideal illustrations of the changes in methodology that have arisen from evolutionary insight.

The nature of the homological relationship was immediately clarified by the discovery of evolution. All that was necessary to supply terms for the relation was to make descent from a common ancestor a defining property for the correspondence. Entities are homologous if, in principle, they can be traced back to a single genealogical precursor. Entities are analogous when, although they have certain similarities, it is not necessarily the case that they can be so derived. (Usage here is ambiguous—analogy is sometimes thought of as including homology, sometimes as excluding it.) The relationships involved are such that they readily lead to verbal confusion. For example, the wings of birds and of bats have evolved separately, but they can be traced back to an ancestral forelimb which was not a wing: as wings, they are analogous; as anterior appendages, they are homologous. For these and more subtle, metaphysical reasons, the word "homology" remains a bone of contention in philosophical biology. Platonists still insist that the archetype must be a defining characteristic of the relation.[15] Nominalists frequently confound definition of the word with evidence for its correct application. Others have attempted to redefine "homology" so as to make it an intrinsic property rather than a relational one.[16] It is revealing that Darwin, who, in his work on orchids, treats the concept of homology at some length, successfully avoids the blunders to which lesser philosophical intellects have fallen prey.[17] He did not, like Owen, insist on calling "the same" organs by definition homologous without defining "the same." Nor did he invoke Platonic essences or a psychologistic "resemblance." His successful dealing with such abstract terminology again suggests that he was a better philosopher—or, if it be preferred, logician—than has been recognized.

A system of homological plans is readily made to serve as evidence for genealogical relationships. Darwin shows how, in different groups of barnacles, a fundamentally equivalent pattern in the external covering is modified in different ways. He demonstrates that in certain forms some of the plates have been fused or altered, although relative positions remain unchanged. He derives the system of plates in the more advanced, sessile barnacles through a series of stages from pedunculated forms, ultimately creating a unitary system of comparisons by which the plates may be traced in their modifications throughout the group.

If one takes a barnacle and pries apart the movable plates which form a sort of trap-door arrangement, one reveals a cavity which encloses such structures as mouthparts, external genitalia, and legs. A barnacle obtains its food by opening up its valves and extending its appendages, rapidly sweeping the adjacent area for any food the water may contain. By the time Darwin was at work on the group, it had become clear that a barnacle is a curiously deviant crustacean, as an adult cemented to some surface, with external plates quite different from anything occurring in other crustaceans, and so peculiar in structure as to render the identification of parts most uncertain. The legs are jointed, like those of other arthropods; this clearly separates barnacles from the bivalved mollusks to which Cuvier had compared them. The discovery by Vaughn Thompson that the larvae of barnacles resemble those of other crustaceans had placed them in closer systematic proximity to crabs and copepods.[18] But the details of morphological correspondence had not been worked out, and the production of barnacles from quite different precursors implied that the entire structure of the body had been greatly reorganized. It is hardly remarkable that Darwin should have been fascinated by the types of modification that had taken place. Swallowed up by a flood of new questions as to homologies, he could not have been satisfied as to the truth of his insights until he had studied the entire group.

Darwin traced the sequences of modification for every structure to which his fairly crude instruments gave him access. He diagrams (fig. 5) a stomatopod crustacean, with its long neck and well-developed thorax, placing beside it a lepadid barnacle, orientated in the same way.[19] He shows how, in the transition from a free-swimming larval stage, the young barnacle attaches by its anterior portion; how the head-section hypertrophies; how plates develop at the side of the body to form a covering; and how the growing barnacle gradually changes in proportions to assume the final, quite different form. He traces out the homologies of the

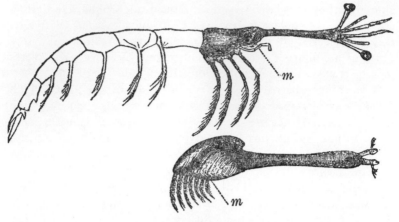

[*m.*—Mouth]

FIG. 5 Comparison of a stomatopod crustacean (above) and an idealized
lepadid barnacle (below) with antennae and eyes (actually lost in
the adult) shown for clarity. From Darwin's barnacle monograph.

legs and shows that these become modified according to different
patterns in the separate groups. And he does much the same for
mouthparts, nervous system, and other structures.

Thus far, we have been dealing mainly with "conventional"
comparative anatomy, based largely on formal properties. Many
analogies may be drawn between this kind of work and stratigraphy. The close parallel between the two disciplines makes it
seem odd that anybody would attempt to elaborate a "pure"
morphology, without historical or causal implications. For the
overwhelming significance of laws of nature in verifying geological theories, and in relating them to an explanatory system, makes
an idealistic stratigraphy—analogous to the idealistic morphology
that still prevails in some quarters—too ridiculous to take seriously. But the belief that life is more than something which certain chemicals happen to be doing was reasonable in the nineteenth century. Only when there was some understanding of the
causal factors involved could comparative anatomy shift its
emphasis from Platonic essences and formal properties to laws of
nature, efficient causes, and historical explanation. In Darwin's
discussions on reproductive structures we may see how the new
orientation in theoretical outlook begins to manifest itself. It is

clear, for instance, that a more functional and explanatory approach underlies his remarks on the cement-glands, structures which aid in attaching the cirripede to its substrate:

> I feel an entire conviction, from what I have repeatedly seen in several genera of the Lepadidae, both in their nature and pupal condition, and from what I have seen in Proteolepas, that the cement-glands and ducts are continuous with and actually a part of an ovarian tube, in a modified condition; and that the cellular matter which, in one part, goes to the formation of ova or new beings, in the other and more modified part, goes to the formation of the cementing tissue. To conclude with an hypothesis,— those naturalists who believe that all gaps in the chain of nature would be filled up, if the structure of every extinct and existing creature were known, will readily admit, that Cirripedes were once separated by scarcely sensible intervals from some other, now unknown, Crustaceans. Should these intervening forms ever be discovered, I imagine they would prove to be Crustaceans, of not very low rank, with their oviducts opening at or near their second pair of antennae, and that their ova escaped, at a period of exuviation, invested with an adhesive substance or tissue, which served to cement them, together, probably, with the exuviae of the parent, to a supporting surface. In Cirripedes, we may suppose the cementing apparatus to have been retained; the parent herself, instead of the exuviae, being cemented down, whereas the ova have come to escape by a new and anomalous course.[20]

Elsewhere, he again refers to the same phenomenon: "I may here venture to quote the substance of a remark made by Professor Owen, when I communicated to him the foregoing facts, namely, that there was a new problem to solve,—new work to perform,— to attach permanently a crustacean to a foreign body; and that hence no one could, *a priori*, tell by what singular and novel means this would be effected." [21] Even though Darwin here takes pains to cast his statements in pre-evolutionary language, he obviously had in mind a purely evolutionary interpretation: there had been a shift in function. The theory of natural selection implies that a structure will not evolve because of some future advantage it might confer upon subsequent generations. Hence, not only must a feature arise out of a preexisting one, but at each stage in its origin it must perform some function, or at least it must exist because of some efficient cause. Further, the acquisition of the new function cannot occur unless, by accident, the ancestral structure possesses those physical and chemical properties which are necessary for the new role. Thus, not only may innovations arise from quite unexpected sources, but also the particular origin may be interpreted as owing to potentialities which already existed in the precursor. Darwin held that a glue-like substance, one which must already have been present in the female tract, happened to be suited to attaching the barnacle to its substrate, and therefore came to subserve the new function. His ability to arrange the various forms in a sequence in which such a modification would have taken place could be invoked as evidence for the hypothesized homology, because the shift in function depends upon a whole series of coincidences which would not have happened unless there had been a particular sequence of changes. Darwin maintained, for instance, that the ancestral barnacle was a hermaphrodite; the use of a female secretion as a means of attachment would have created great difficulties in forms with separate sexes, for the males would have had no such means of attachment. Thus it would appear that Darwin tested at least some of his comparative anatomical ideas by reconstructing the sequence of historical events responsible for present attributes and seeing if such sequences were intelligible as the result of natural selection. In so doing, he was providing critical tests of his hypotheses, both systematic and evolutionary. Hence it is demonstrable that our knowledge of past evolutionary events need not be based solely on the enumeration and correlation of similarities and differences. Logically, our knowledge of phylogeny derives from a model, verified by reference to laws of nature and the ability to predict the attributes of organisms.

One might argue that consistency with evolutionary theory does not prove a phylogenetic hypothesis, because a lack of a contradiction might have a variety of meanings. This view is tenable, although it is irrelevant to the validity of Darwin's reasoning. Scientific theories cannot be proved to be correct; they can only be refuted. The attempt to see if a phylogenetic theory is consistent with our understanding of evolutionary mechanisms is basically an attempt to falsify hypotheses. One can never be sure that some future observation will not prove inconsistent with the hypothesis. Although Darwin's reasoning was both valid and original, his views on the homologies of the cement glands have had to be considerably altered in the light of new knowledge.[22] Likewise, the basic premise that the ancestral barnacle was hermaphroditic can be challenged.

Darwin's manner of dealing with problems related to the loss of hermaphroditism provides another example of his method, but this time one which has more successfully withstood criticism in the light of new facts. The barnacles generally available to Darwin were hermaphrodites, although several were gonochoric (i.e., separate-sexed). One naturally wonders which condition came first. Did hermaphrodites give rise to gonochorists? Or did a lineage in which the sexes are separate give rise to forms in which the male and female organs are united in the same individual? The conditions of sexuality in barnacles are rather odd: Where males are present, they generally are small, and live attached to the females. These so-called "dwarf-males" are not unique to barnacles, being known from a number of animal groups. Darwin came across an intermediate condition—a hermaphrodite with a dwarf male. In a letter to Hooker, written in 1848, he recounts this discovery:

> I have lately got a bisexual cirripede, the male being microscopically small and parasitic within the sack of the female. I tell you this to boast of my species theory, for the nearest closely allied genus to it is, as usual, hermaphrodite, but I had observed some minute parasites adhering to it, and these parasites I now can show are supplemental males, the male organs in the hermaphrodite being unusually small, though perfect and containing zoosperms: so we have almost a polygamous animal, simple females alone being wanting. I never should have made this out, had not my species theory convinced me, that an hermaphrodite species must pass into a bisexual species by insensibly small stages; and here we have it, for the male organs in the hermaphrodite are beginning to fail, and independent males ready formed. But I can hardly explain what I mean, and you will perhaps wish my barnacles and my species theory al Diavolo together. But I don't care what you say, my species theory is all gospel.[23]

One might ask just why this discovery was so convincing, and just why it was so difficult to explain. Perhaps the species theory was merely of heuristic value—it may have suggested that one should look for intermediate forms between different groups and structures. This interpretation, though a partial clue, is not adequate, in view of the fact that systematists traditionally had looked for interconnecting forms. And it casts no light on the problem of Darwin's views on what came first, the hermaphroditic or the gonochoric state. To be sure, he had the closest relative to compare, but this would give no definitive argument. A closer scrutiny of Darwin's statements, in the light of his other ideas, shows that it was not the intermediate forms as such that impressed him, but the details of structure, interpreted in the light of theory.

It is easy to see how advantageous it would be for a barnacle, or, for that matter, any other sessile organism with internal fertilization, to have some way of increasing the probability of finding a mate.[24] A barnacle may have only its nearest neighbor available for the exchange of sperm. If the sexes occur in a one-to-one ratio, and if they are separate, the chances are equal that both will be of the same sex, and therefore that fertilization will be impossible. Hermaphroditism is one way to overcome this problem, as the barnacle can exchange sperm whenever any fertile member of its species gets within range. Another way is by dwarf males: when a larva settles alone, it turns into a female; if it happens to meet up with a female, it turns into a male and the two live in close association. Or perhaps the females alone will settle in isolation, and the males only on females. Whichever may be the case, a female barnacle would be almost assured of a mate, as settling sites are far less numerous than the larvae available to occupy them. With two solutions available for the same problem, it is clear that there is no way, merely on the basis of formal properties, to tell which stage came first. A gonochoric form could give rise either to hermaphroditic forms or to ones with dwarf males. Alternatively the forms with separate sexes and dwarf males could give rise to hermaphrodites, or vice versa. Evidently Darwin answered the question of which came first through the analysis of the possible ways in which the change might occur. He reasoned that a hermaphrodite changing into a gonochorist would pass through a stage in which the male organs were still present, but were superfluous, owing to the presence of the dwarf males. And the presence of small, male structures in the hermaphrodite could best be explained if these were in the process of reduction. One would not expect the gradual development, by natural selection, of a complicated set of reproductive organs, through stages in which they were useless, then superfluous, "for the purpose" of subserving a definite function millions of years in the future. Darwin had discovered instances of what is called rudimentary hermaphroditism, a condition that has arisen independently in a number of primarily hermaphrodite groups (such as trematodes), where the males and females in the derived forms have come to live in close association, and where hermaphroditism is no longer advantageous. Thus, the direction of change was derived by considering the kinds of selection pressures that must have existed under the prevailing conditions of life. Hypotheses about past evolutionary events were conjoined with equally hypothetical laws of nature, and both were tested by observations on the organisms themselves. In effect, Darwin had devised a comparative anatomical method based on the theory of natural selection and involving the functional analysis of vestigial structures. In pre-Darwinian comparative anatomy, whether evolutionary or not, arguments of this type did not exist. The difference is clear from a paper written by Wallace at a time when he had embraced evolution but had not yet discovered its mechanism: he thought that vestigial organs were *in statu nascendi*.[25] Darwin's use of vestigial organs as an argument against special creation is well known; his taxonomic applications of the same principle are a bit too obscure to have been generally appreciated.

But Darwin's views on the sexuality of cirripedes are far more involved than this simple example, or even than the barnacle monograph as a whole, is likely to suggest. In fact, his interpretation of the sexual relations of barnacles is an integral part of a comprehensive theory of sexual biology, one which may be traced back to his early speculations on evolution, and which is fundamental to several of his major works. His taxonomy and anatomy of barnacles provide striking verification—and application—of this theoretical system. It is profoundly revealing that Darwin worked out the major portion of his sexual biology "deductively," before undertaking his studies on barnacles and on the reproductive physiology of plants. In his fourth notebook on transmutation of species, he observes that the cirripede sexual system is remarkable in view of the fact that most arthropods have separate sexes.[26] He goes on to develop the theory, one of his major contributions to knowledge, that hermaphrodites generally cross-fertilize, a view he later verified in his work on barnacles and in his experimental studies on plants. He then elaborates upon the close correlation between sessility and hermaphroditism, drawing an analogy with plants. In the barnacle monograph, he returns to this analogy: the complemental males, hermaphrodites, and forms with separate sexes all find parallels in various groups of flowering plants.[27] It seemed that the type of sexuality had become modified repeatedly in different lineages.

In his notebooks Darwin also expresses the beginnings of a theory that every animal is basically a hermaphrodite.[28] In *The Descent of Man* he once again takes up this topic in a discussion on vertebrates.[29] He cites the authority of the German anatomist Gegenbaur for this theory, and argues for it on the grounds that "one sex bears rudiments of various accessory parts, appertaining to the reproductive system, which properly belong to the opposite sex: and it now has been ascertained that at a very early period both sexes possess true male and female glands." [30] Why there should be rudiments of this nature, such as the nipples on a man's chest, was a most difficult question at the time, and remains obscure in some details. Darwin rejects the view that such accessory structures of mammals evolved at a time when mammals were hermaphrodites, on the grounds that separate sexes are the universal rule in higher vertebrates. His explanation, and it still stands, is that such organs have been "gradually acquired by one sex, and then transmitted in a more or less imperfect state to the other." [31] Here Darwin is invoking a "pleiotropic effect," a concept which he applies repeatedly in his works. If a character, selected because of its utility for one sex, happens to be produced by a morphogenetic system which affects the other sex in the same way, that character will be present in both sexes. It is now known to be a generally applicable rule that any individual possesses, potentially, the necessary mechanisms for producing the phenotype of either sex.[32] This was inferred by Darwin through such evidence as observations on the effects of aging and castration in fowl, phenomena which had been under scientific investigation since the founding of endocrinology early in the century.

One possible reason why organisms are potentially capable of developing the characteristics of the other sex is that neither sex can develop without the interaction of both male-forming and female-forming substances. Darwin evidently held that the association began at a stage when both male and female systems were active in the same individual. Certainly, this would explain the facts, but other possibilities must be considered. Darwin would appear to have differentiated, but perhaps not well enough, between two kinds of homology, which are not always clearly distinguished. In addition to evolutionary homology, there is another class of homologies, generally treated as several. These include: (1) serial homology, such relations as exist between the human femur and humerus, that is, the "same" parts in different sets of structures repeated in a linear series; (2) the relation between the "same" joint in all fingers of one hand; (3) the relation between the comparable parts on the right and left sides of the body; and (4) the relation between comparable parts in different sexes (which will here be called *sexual homology*). Most of these correspondences exist because similar structures may be produced in a single individual through a repetition, perhaps with variations, of a single kind of developmental process. Different sexual forms within a species could likewise arise through modifications in the ontogeny of a single kind of system. In the sixth edition of *The Origin of Species* Darwin credits Lankester for such distinctions.[33] However, it is clear not only that Darwin understood the problems, but also that he was the first to relate such phenomena as serial metamerism to the underlying causes. He did not blunder into the use of such terms as "homoilogy," "homoplasy," and so forth, which result from ignoring terms and from trying to make relational properties intrinsic. Serial and sexual homologies are comparable to the evolutionary ones in the sense that the observable correspondences in formal properties are due to the presence of the "same" morphogenetic process; but the word "same" has quite different meanings. The several structures produced by the action of the "same" morphogenetic process do not evolve independently of each other, as do evolutionary homologues; they tend to respond in the same way to modifications of the developmental system, through what Darwin called "correlation of growth." Mammalian teeth, for example, sometimes respond as a unit during evolutionary change; hence a character present in all teeth could well have originated in only some of them.[34] Nor do we infer that the ancestral vertebrate had only one tooth. On the same grounds, an originally hermaphroditic state in animals is not supported by the sexual homology, but is only consistent with it.

It would seem that Darwin had another reason for considering the hermaphroditic condition to be the primitive one. His opinion may have arisen from a consideration of how one state could lead to another. Darwin does not state his precise reasons explicitly, but we can suggest the type of argument he might have used.[35] It is hard to imagine how, in an organism with so complicated a reproductive system as a higher vertebrate, a system with separate sexes could have given rise to a hermaphroditic one. Were the two systems to be superimposed upon each other suddenly, the sexually homologous structures would probably be rendered inoperative, since they subserve quite different functions in male and female, differ greatly, and are delicately adjusted. It is known that intersexes are generally sterile in higher organisms, although in such animals as clams, where the reproductive system is very simple, normally gonochoric species may produce functional hermaphrodites. The hermaphroditic state, where it has evolved from the gonochoric, would appear to have been derived through an intermediate condition of sequential hermaphroditism. That is, the animal begins life as male or female, but changes sex. The gastropods would seem to have undergone such a sequence of modifications. The "lowest" snails have separate sexes, with a labile system of sex determination which readily allows organisms of one sex to be converted into the other. Many forms of intermediate grade are sequential hermaphrodites. From these latter, evolutionary change would appear to have gone in two directions, paralleled by other modifications in the reproductive system: one group is strictly gonochoric, while the other has tended toward simultaneous hermaphroditism, with both sexes functional at the same time.[36] Thus it is a fairly complicated and difficult process through which a gonochorist may be evolved into a hermaphrodite. But consider the opposite transformation. If one starts out with a simultaneous hermaphrodite, in which the genital structures of both sexes coexist in a functional state, all that is necessary to separate the sexes is a division into two groups, each carrying out the functions of only one sex.

Further evolution toward gonochorism could proceed simply by the selection, in each sex, of genes which make the structures of the other sex fail to develop. In a complicated reproductive system, such a simplification could be most advantageous if, when coexisting in the same individual, the male and female structures tended to interfere with one another. Such changes are known to have occurred in various hermaphroditic groups, such as gastropods. Evidently Darwin thought that a separation of the sexes had occurred in barnacles, and in other groups as well, through a repression of the developmental processes which give rise to the structures proper to each sex.[37]

This brings us to another of Darwin's basic theoretical systems, the one concerned with the relationships between evolution and ontogenetic development. As Darwin never wrote a treatise on physiological embryology, this particular aspect of his thinking has been almost completely overlooked. Nonetheless, it pervades his reasoning in all of his biological works, and to ignore it has been to disregard a fundamental premise, and thereby to render much of his work unintelligible. Darwin's own views on the relationships between development and evolution are obscured by the existence of certain other interpretations. During the pre-Darwinian period a number of biologists of no mean competence—Meckel, Serres, von Baer, Agassiz, and others—had been impressed by the fact that there is some degree of correspondence between the stages of individual development (ontogeny) and the groupings of the taxonomic hierarchy. Such a phenomenon can have any number of possible meanings and may even be given a metaphysical interpretation. To a Platonist, for instance, it could correspond to the manifestation of a system of Ideas having existence in the mind of God. Or, if the Platonist were to advocate evolution, it could mean that the Ideas gradually manifest themselves in the development of both individuals and lineages—a curious variant on the traditional analogy between microcosm and macrocosm.

To a mechanistic evolutionist, the correlation between ontogeny and the taxonomic hierarchy could be explained in terms of historical relationships: the developmental pattern bears traces of the same course of events that is responsible for our ability to erect the system of classification. Darwin used this correspondence as one of the basic arguments in *The Origin of Species.*

Soon afterward, Ernst Haeckel (1834–1919) elaborated a remarkably effective propaganda device: the so-called Biogenetic
Law, which states that ontogeny recapitulates phylogeny. Few
scientific ideas have been responsible for so much confused and
vituperative controversy. The underlying issues are still not generally understood. Haeckel was surely too much of an enthusiast.
Evidently he felt that evolutionary change occurs by the addition
of new stages at the end of development. As a rule admitting of
no exceptions, this premise is false. De Beer, for example, has
labored mightily to adduce evidence (of varying reliability)
showing that evolutionary novelties may arise at any stage in the
life cycle, or that later stages may be lost.[38] And it is abundantly
clear that the principle of recapitulation has but a limited validity: it is little better than a heuristic aid. But de Beer's assertion
that the exceptions invalidate the generalization is founded on
error. His argument is based on his personal decision to use the
word "law" to designate only those generalizations which admit
of no exceptions. But if one means by "law" a valid generalization
about the behavior of matter, then it is perfectly acceptable to
refer to recapitulation as a "law," as long as the usage is made
clear. Furthermore, all laws admit of exceptions when the conditions are not met. The historical evidence brought forth by de
Beer to show that von Baer's theory is preferable has been adequately criticized by Lovejoy.[39] It is even more revealing to
compare Haeckel's own words with what his various critics
allege that he said. Sedgwick asserted that there must be a perfect
correspondence between the developing embryo and the adult
ancestor.[40] Haeckel, on the other hand, writing in the same
anthology, says: "Ontogeny (embryology or the development of
the individual) is a concise and compressed recapitulation of
phylogeny (the paleontological or genealogical series) conditioned by laws of heredity and adaptation." [41] It may seem
curious that both Sedgwick and de Beer ignore Haeckel's qualifications, which amount to a statement of conditions which cover
the exceptions. This is typical: as with Darwin's coral reef theory,
the issue became an emotional and a metaphysical one. Likewise,
what Haeckel meant by phylogeny being the "cause" of ontogeny has been grossly misinterpreted. The word "cause" has a
number of meanings, and it is no simple matter to exclude the
possibility that one of them would be valid. Such vagueness and

uncertainty provide most sensible grounds for abandoning the biogenetic law as having little utility. The real reason why Haeckel has been so unjustly criticized, however, has little to do with the truth or falsehood of his biological theories. Haeckel has been persecuted because he was a popular leader in the nineteenth century struggle against Prussian despotism. He was a materialistic pantheist (as he put it, a "monist"), who used evolution as a means of attacking the established order. His battle with Virchow over the teaching of evolution in the schools of Germany was an episode in the long conflict between Platonism, aristocratic privileges, and ecclesiastical tyranny on the one hand, and empiricism, popular democracy, and freedom of conscience on the other. Haeckel has a bad reputation largely because people have believed what reactionaries have written about him.[42] Of course, another kind of politics—right, left, or center—means another kind of folly, but this only reinforces the obvious inference that it is all too easy for a scientist to be cast adrift on a sea of metaphysics.

Darwin's views on recapitulation are formulated at some length in *The Origin of Species*.[43] He argues that variations may occur at any stage in the life cycle. At whatever stage variation does take place, it will affect the properties of all the following stages, but not the earlier ones. A lack of either selection pressure or variation in early stages should result in the young being less modified than the adults. It is true that Darwin altered later versions of *The Origin of Species* to accommodate Haeckel's idea that the young stages might in some way correspond to adult ancestors.[44] However, the degree and kind of such correspondence—precisely which characters are larval adaptations and which reflect the structure of adult ancestors—are not discussed. There is no reason to think that Darwin ever substantially altered his views from those expressed in the "Sketch" of 1842: "It is not true that one passes through the form of a lower group, though no doubt fish [are] more nearly related to [the] foetal state. (They pass through [the] same phases, but some, generally called the higher groups, are further metamorphosed.)"[45]

Such an interpretation of Darwin's views on recapitulation may be supported by demonstrating that it is implicit in a more general conception of developmental mechanisms, from which he drew numerous conclusions. It seems that Darwin conceived of morphogenesis as an integrated process of growth and change, and of variation as resulting from alteration in the rate or quality of the developmental process. Were one to ask what would be the effect of altering a series of morphogenetic changes at various times in the life cycle, the obvious answer would be that, on the whole, young stages should become altered less rapidly than adult ones. And the exceptions could be and were shown to have resulted from more intense selection pressures having acted on the younger stages. In other words, Darwin constructed an abstract "model" of developmental mechanisms generally and saw that, when combined with his ideas on natural selection, they accounted for the correspondences between ontogeny and phylogeny. This gave rise to a theory which was more than a mere statement of an observed correlation. Nor was Sedgwick right in saying that "recapitulation theory is itself a deduction from the theory of evolution." [46] Both evolution and recapitulation are "deductions" from hypotheses about embryology and natural selection, and inductions justified by the success of experiential testing.

That Darwin actually did use such a model of embryological mechanisms can be verified by showing that he drew other conclusions from it. In subsequent chapters concerned with *The Variation of Animals and Plants Under Domestication*, with plant physiology, and even with behavior, it will be argued that many of Darwin's statements are intelligible in the light of this hypothesis. The work on barnacles is likewise a source of examples which can illustrate this approach. Many of Darwin's statements are here cast in morphogenetic terminology. Pre-evolutionary anatomy often compared organisms in terms of embryological changes, in which the degree of perfection attained was thought of as corresponding to the extent to which the archetype had been manifested. Hence it was possible for Darwin to compare barnacles by referring to the degree of development in parts and characteristics. No change in the wording is necessary to give his statements an evolutionary meaning; and in retrospect it is obvious that Darwin intended it that way. Central to his reasoning in much of the work is the concept of *developmental arrest*. In *The Variation of Animals and Plants Under Domestication*, and elsewhere as well, Darwin points out that many monstrosities were known to result from the termination of developmental processes. He enumerates many examples of monstrosities bearing a close resemblance to earlier stages in normal development, citing the works of Isidore Geoffroy Saint-Hilaire (1805–1861). It is particularly noteworthy that Isidore's father, Étienne Geoffroy Saint-Hilaire (1772–1844), who has already been mentioned in relation to the debate with Cuvier, was an early evolutionist who also approached the problem of evolution from a developmental point of view. Cahn has lately drawn attention to an obscure work by the elder Geoffroy, in which a parallel is drawn between the sexually mature, but morphologically larval, salamanders which inhabit certain caves, and the experiments of William Edwards showing that tadpoles kept in the dark do not metamorphose.[47] Geoffroy held that there was a cause and effect relationship here—the environment producing an adaptive change. We can now link up both phenomena to an underlying process: the effect of light on the production of hormones. The evolution of sexually mature larval forms, with the consequent elimination of adult stages, is called *neoteny*. It is a fairly common phenomenon, and one which de Beer invokes as contradicting recapitulation. De Beer denies that Darwin knew of neoteny; yet in one of his notebooks, Darwin draws the same parallel between cave salamanders and tadpoles as did Geoffroy, even mentioning Edwards.[48]

Darwin also discovered neoteny in barnacles. Indeed, not only did he discover it, he verified it, explained its relationship to developmental physiology, and worked out techniques for overcoming the difficulties it poses for systematics. In *The Origin of Species* his reference to "retrograde" development of the complemental males is a bit obscure, but his meaning is clear in the original monograph.[49] We have already noted his assertion about the male organs "beginning to fail" in a hermaphroditic barnacle. He observes that when the sexes are separate, the males are small and epizoic, pointing out how they resemble immature forms and referring to the "abortion" of certain segments.[50] *Analesma* is said to be "in some degree in an embryonic condition." [51] Observing that the dwarf male of *Ibla* resembles an embryonic female, he comments on the ecological parallels in other groups: "It deserves notice that in the class Crustacea, both in the Lerneidae and in the Cirripedia, the males more closely resemble the larvae, than the females; whereas amongst insects, as in the case of the glow-worm in Coleoptera, and of certain nocturnal Lepidoptera, it is the female which retains an embryonic character, being worm-like or caterpillar-like, without wings. But in all these cases, the male is more locomotive than the female." [52] Even more striking is the following: "It is very singular how much more some of the Males and Complemental Males in Scalpellum differ from each other, than do the female and hermaphrodite forms; this seems due to the different stages of embryonic development at which the males have been arrested." [53] This passage is particularly significant, not only because it shows that Darwin looked upon males as having beeen produced from hermaphrodites when certain structures ceased to develop, but also because it reveals the role of rational—as opposed to intuitive—interpretation of taxonomic evidence. His classification was derived not from the number of shared attributes, but from the manner in which these attributes relate to laws of nature and to historical hypotheses. He treated his ideas about phylogeny and embryology as theories and tested both of them by their joint implications. The same type of method is applied where he finds a sessile cirripede superficially resembling a pedunculated form: he reasons, from functional considerations, which similarities should be due merely to adaptation to the same way of life, and which should be used as reliable evidence for relationships. [54] And he gives considerable weight to the comparison of antennae, on account of there being no reason to expect these larval parts to have been subject to the selection pressures which affect adults. [55] This process of evaluating the evidence, which has been misleadingly called weighting of characters, is a standard procedure in the work of comparative anatomists—Cuvier was famous for his skill in doing this through functional considerations. It is an application of the reasonably obvious fact that laws of nature affect the visible properties of organisms. By formulating and verifying hypotheses about the effects of such laws under particular conditions, one may distinguish that which is striking to the eye from that which is relevant to the understanding of past events. The operation of weighting is quite straightforward: it scarcely differs from what a jury does when faced with conflicting testimony. The very possibility of such an operation has been denied by some writers, in spite of common sense, and in spite of the fact that comparative anatomists have been doing it successfully for years. But in view of the many examples adduced so far, it should be obvious that such confusion is only to be expected, and that its explanation is to be found not in science, but in metaphysics.

Darwin reasons from the modification of developmental processes to explain, and to investigate, far more than just the origin of dwarf males. *Analesma* is interpreted as being embryonic as an adult.[56] In *Alcippe* (= *Trypetesa*), he submits the body as a whole to a comparison based on morphogenetic concepts.[57] He notes that in adult *Alcippe* both sexes, like many larvae, lack an anus, and compares them to the larvae of certain parasitic wasps.[58] Most remarkable of all, he observes how the body may be treated as having undergone, along a sort of gradient, an acceleration of growth at the anterior end, while the posterior portion remains in an embryonic condition.[59] In another form, he not only shows how rudimentation proceeds along an anterior-posterior gradient, but draws an analogy with plants:

> In the male *Ibla*, abortion has been carried to an extraordinary and, I should think, almost unparalleled extent. Of the twenty-one segments believed to be normally present in every Crustacean, or of the seventeen known to be present in Cirripedes, the three anterior segments are here well developed, forming the peduncle: the mouth consists as usual of three small segments: the succeeding eight segments are represented by the rudimentary and functionless thorax, supporting only two pairs of distorted, rudimentary and functionless cirri: the seven segments of the abdomen have disappeared, with the exception of the excessively minute caudal appendages; so that, of the twenty-one normal segments, fifteen are more or less aborted. The state of the cirri is curious, and may be compared to that of the anthers in a semi-double flower; for they are not simply rudimentary in size and function, but they are monstrous and generally do not even correspond on opposite sides of the same individual.[60]

This approach anticipates by many years the geometric treatment developed by D'Arcy Thompson, involving the construction of graphs depicting organisms which are deformed according to an equation.[61] It is abundantly clear from Darwin's quantitative, although not geometrical, treatment of the same type of relation-

ship in *The Variation of Animals and Plants Under Domestication,* that he fully understood the principle, and that he related it
to a more general theory of embryology.[62] Others have attempted, with some success, to apply such geometrical techniques
to systematics, but largely without carrying the analysis beyond
the demonstration of a formal equivalence.[63] Darwin did not
invent the geometrical approach—it was developed at great
length by such artists as Albrecht Dürer (sixteenth century).[64]
But in relating the principle to underlying causes, and in applying
it to taxonomy, Darwin manifests his prowess as both innovator
and theoretician. To be sure, his cleverness is not particularly
obvious until we place the deformation of the body and the loss
of the anus in relation to the concept of the correlation of
growth: that is, to the idea, basic to so many of Darwin's
arguments, that different parts of the body may evolve as a unit.
Selection for one property tends to affect all those structures
which are bound up with the same developmental mechanism,
with the consequence that adaptively disadvantageous attributes
—as in the loss of the anus—are evolved. Once again, we see that
the truth of Darwin's taxonomy follows logically, not simply
from classes of similar entities having been grouped together, but
from the manner in which the facts stand in harmonious relation
to laws of nature, in this case those of developmental mechanics.
The subjective apprehension of resemblances has no more weight
in an argument for a scientific taxonomy than does a quotation
from Scripture.

PLACE OF THE BARNACLE MONOGRAPH IN
DARWIN'S SYSTEM

In view of the logical structure of the barnacle monograph, it is not difficult to see why Darwin revised the entire
subclass of cirripedes. The work is a system of interlocking concepts, depending for its justification upon the coherence of the
argument as a whole. It embraces several intermeshing theories.
One is the theory of natural selection, which implies that sequential changes must be functional at all stages. Another is his morphogenetic system, determining the meaning and relevance of
many otherwise incomprehensible correlations. An intricate

theory of sexuality is fundamental to much of the argument. One may likewise find traces of paleontological, biogeographical, and even behavioral reasoning. A revision of the entire group, demonstrating the gradual passage of one form into another, and with the groups, from species upward, reliably delimited, was essential for a sound interpretation of the more fundamental biological theories. For only when the whole system was finished could it be seen that each of the separate trains of thought fit in with all the others. The completed work was nothing less than a rigorous and sweeping critical test for a comprehensive theory of evolutionary biology.

The *Monograph on the Sub-class Cirripedia* manifests the pattern typical of Darwin's works in general. His notebooks demonstrate that, as in the work on coral reefs, many of the theoretical elements were elaborated "deductively," before the gathering of new observations began. And it is marked by the same thoroughness of investigation: the devotion of eight years to cirripedes is readily compared to reading "every work on the islands of the Pacific." [65] Darwin's ability to use systematics as a means of verifying hypotheses about the nature of organic processes is of the greatest importance to modern science, for the literature abounds in assertions that it is impossible to derive, through taxonomic study, such knowledge of phylogeny and evolutionary mechanisms as Darwin did in fact obtain.[66] Those who make such pronouncements habitually argue from the premise that they, personally, cannot see how it could be done. Their argument combines the modesty of Schopenhauer with the logic of Mary Baker Eddy; it does not follow from their own lack of imagination that Darwin or anyone else must fail.

As in the rest of Darwin's works, the excellence of the barnacle mongraph lies, not simply with the facts it unearths, but with the underlying logic of the system. Although many of the elements in this, as in the rest of his works, may be found in the thought of his predecessors, these gain an altogether different significance in the finished structure. And the replacement of an "inductive" and formalistic morphology by a hypothetico-deductive and evolutionary comparative anatomy ushers in a major revolution in the study of organisms. Darwin's genius is not immediately manifest, for he deals with intricate problems; and his thinking is not to be understood without some knowledge of the animals themselves. Yet when the full meaning of the work is grasped, the *Monograph*, like the *De Motu Cordis* of William Harvey, takes on the deepest interest as an innovation in comparative anatomical methodology.[67]

6. A Metaphysical Satire

Several of Darwin's works become far more meaningful when related to the comprehensive system of sexual biology which had such an important role in his barnacle studies, and which will take on renewed significance in the analysis of *The Descent of Man*. Sex is so familiar that we tend to overlook its remarkable aspects. Why, for example, should organisms reproduce sexually at all? Many organisms survive for generations by asexual propagation: budding, regeneration, or fission. We tend to lose sight of the fact that prior to Darwin no real answer to this question was available; indeed, our understanding is far from complete. Our appreciation of sex has depended largely upon insights into evolutionary mechanisms. The common occurrence of sexuality, especially in plants and microorganisms, was discovered more recently than one might have expected, reflecting the general rule that theory plays a major role in determining what scientists will observe.

THE EVOLUTIONARY INTERPRETATION OF SEX

The development of Darwin's sexual theories may be traced through progressive stages in his works. As usual the investigation began with theoretical considerations. Darwin's first notebook on transmutation of species refers to the *Zoonomia* of his grandfather Erasmus Darwin, a medical doctor, philosophical poet, and early advocate and popularizer of evolutionary ideas.[1] Erasmus Darwin was the author of such curiosities as *The Temple of Nature* and *The Loves of the Plants,* and it was perhaps from him that Charles Darwin obtained the idea that sex plays a significant role in evolutionary processes. There is an obvious, but hard to interpret, relationship between sexual reproduction and change in character, in that organisms propagated by grafting or other kinds of asexual reproduction more closely resemble their parents than do those produced by sexual processes. But the nature of this relationship was not at all clear in Darwin's mind until after he had read Malthus and had formulated the hypothesis of natural selection. In his first notebook, he takes over his grandfather's view, saying that generation is "a means to vary or adaptation." [2] In the second, he asserts that the "final cause" of there being two sexes is "adaptation of species to circumstances." [3] At last, early in the fourth notebook, writing after he had read Malthus, he states: "My theory gives great final cause of sexes in separate animals: for otherwise there would be as many species, as individuals. . . ." [4] This was a most important point, for he was in fact recognizing the fundamental role of sexual reproduction in giving functional unity to reproductive populations. Sex distributes an emerging property throughout a group. It thus brings about a change in character of the population, as distinguished from a change in an individual. Darwin considered sex a *sine qua non* for evolution, as may be seen from his assertion that "My theory only requires that organic beings propagated by gemmation do not now undergo metamorphosis, but to arrive at their present structure they must have been propagated by sexual commerce." [5] This inference holds up quite well today. Although it is true that asexually reproducing organisms can evolve by the survival of mutant clones, this kind of evolution is largely restricted to microorganisms, which reproduce very rapidly. For the most part, the abandonment of sexuality leads to an evolutionary dead end, for it does away with the advantages that result from recombination. That Darwin came to this conclusion only after he had read Malthus corroborates once more the hypothesis proposed in chapter 3 that there was a shift in Darwin's thinking from group selection to the more limited form of natural selection; and we have further grounds for rejecting many criticisms of Darwin's thought.[6] It is thus readily intelligible why Darwin, in revising the first, pre-Malthusian edition of the *Naturalist's Voyage*, deleted the following passage:

> We may consider the polypi in a zoophyte, or the buds in a tree, as cases where the division of the individual has not been completely effected. In this kind of generation,

the individuals seem produced only with relation to the present time; their numbers are multiplied, but their life is not extended beyond a fixed period. By the other, and more artificial kind, through intermediate steps or ovules, the relation is kept up through successive ages. By the latter method many peculiarities, which are transmitted by the former, are obliterated, and the character of the species is limited; while on the other hand, certain pecularities (doubtless adaptations) become hereditary and form races. We may fancy that in these two circumstances we see a step towards the final cause of the shortness of life.[7]

It would seem that Darwin emphasized the importance of sex in giving unity to the population, at the price of overlooking its importance in providing variations upon which selection might act. Olby goes so far as to suggest that this point of view caused Darwin to embrace a number of erroneous notions about heredity, especially the concept of blending inheritance.[8] Although there may be some truth in Olby's interpretation, it will be shown in a later chapter that Darwin's genetics may be explained more effectively on the basis of other considerations. If Darwin held that sex is absolutely indispensable for evolution, it would be implicit in his theoretical system that sexual reproduction is far more common than had been thought. Further—and this is a point that group selectionists overlook—there would have to be some reason why sexual reproduction arose and is maintained. For to a group selectionist, but not to Darwin, species which are sexual survive because of a long-term tendency to endure, owing to their evolutionary plasticity. But Darwin had to seek some other reason for the origin and maintenance of sexuality. He was only partially successful in this endeavor, though he did show, through a substantial body of comparative and experimental work, that crossing did appear to have previously unsuspected effects; and many of his incidental discoveries proved to have remarkable significance.

Yet another implication of his theory led Darwin to investigate sexual phenomena. The mechanism which produces adaptations is an indirect one: reproductive success alone determines whether a property will be evolved. And the process of evolution is utterly without foresight or design. It follows that under certain conditions properties may develop which, in spite of their efficacy in promoting reproductive success, would have detrimental effects on individual and species alike. One would expect the evolution of all sorts of contraptions or "Rube Goldberg devices," constructed so as to further some aspect of reproduction, but put together in an impractical and, in the long run, deleterious manner. Evolution, proceeding blindly and without reason, adapts only to the needs of the moment, and not very well at that. It somewhat resembles the proverbial military school which produces officers admirably equipped to fight the battles of previous wars. Such a conception of the evolutionary process leads to a host of readily verified predictions about the properties of organisms, predictions which could not be derived from any of the alternative hypotheses. In treating organisms from a point of view antithetical to design, means were at hand for a devastating argument against the prevailing conceptions of adaptation.

The importance of sexual biology in Darwin's barnacle anatomy has already been discussed. Likewise, it has been demonstrated how the sexual relations of barnacles were made to cast light on evolutionary, embryological, and taxonomic ideas. Indeed, Darwin went so far as to write, in a letter of October, 1852 (seven years before *The Origin of Species* was published): "My first volume is out; the only part worth looking at is on the sexes of Ibla and Scalpellum." [9] So unexpected were the phenomena of complemental males, except as implicit in the theory of natural selection, that in his autobiography Darwin related, with obvious pleasure, that "This latter discovery has at last been fully confirmed; though at one time a German writer was pleased to attribute the whole account to my fertile imagination." [10] Yet a full exposition of Darwin's ideas on the importance of sex had to await the development of his evolutionary theories and a large body of experimental work. In his barnacle monograph, he could do little more than draw some analogies with plants and remark: "I infer that there must be some profounder and more mysterious final cause." [11]

ORCHIDS AND PURPOSE

Darwin must have had the broader, theoretical and philosophical implications of selection theory fully in mind when he wrote his book on orchids. For although the work appears, superficially, to be little more than an entertaining account of some remarkable adaptations subserving fertilization, the work had deeper meanings. In the first place, it illustrated the enormous importance, implicit in Darwin's theory, of cross-fertilization. It could also be read upon another level. The book is a metaphysical tract. It constitutes a deliberate, planned attack upon the argument from design. Even though it is fairly well recognized that Darwin treated the problem of teleology in his writings, the relevance of this aspect of his work to his stature as a thinker has not received due attention. The problems with which he dealt were sufficiently obscure, and his manner of dealing with them was so unconventional, that the profundity of his insight has long eluded the casual reader. Although John Dewey, in a celebrated essay first published in 1910, explained in the clearest terms how Darwin's discovery had overthrown the notion of final causes, and although reputable biologists all but universally condemn teleological thinking, the basic issues remain confused in some circles.[12] Before *The Origin of Species* was published, the major argument for the existence of God was the notion that the universe appears to have been constructed according to some rational plan. The immediate decline of natural theology after 1859 attests to the significance of Darwin's role in bringing about its demise. Modern philosophy would seem quite generally to admit that explanations of vital phenomena in terms of final causes can always be recast into nonteleological terms.[13] Especially in the thinking of biologists, the idea of purpose is looked upon as an unnecessary and delusive anthropomorphism.

Human nature being what it is, so attractive a notion as the argument from design has never really died out, and all sorts of reconciliations have been attempted. On a strictly philosophical plane, for instance, one might argue that simply because teleological explanations are unnecessary, this does not exclude the existence of purpose in nature. True, but neither does the fact that nobody ever saw a unicorn contradict the possibility that someone may. Scepticism of this sort allows the possibility of design, but in no way supports it. The real reason is to be sought elsewhere, and the early history of the arguments that arose out of Darwin's discovery are particularly well suited to show just why the notion of teleology has led to so much delusion.

That Darwin intended his work on orchids as an attack upon

the argument from design can scarcely be contested. In his autobiography he praises the logic of Paley's *Evidences of Christian-
ity* and of other works by the same author which he read as a
student.[14] Paley often speaks of the "contrivances" to be seen in
nature, and argues that the very existence of these necessarily
implies a Contriver.[15] In a work which Darwin read and greatly
praised, the Harvard botanist Asa Gray (1810–1888) persisted in
using the word "contrivance" for "adaptation." [16] Gray, one of
Darwin's major collaborators, was the foremost advocate of Darwinism among American intellectuals in the years immediately
after the publication of *The Origin of Species*. It was he who was
most responsible for winning victory over the Platonism of
Agassiz and obtaining a fair hearing for Darwin's ideas. But although Gray, a devout Christian, accepted evolution and natural
selection, he would not give up design, and he proposed what
Dewey called "design on the installment plan," the idea that the
variations which are selected were specially preordained and
foreseen from the beginning by the Creator.[17] With a fine sense
of irony, Darwin entitled his work of 1862 *"On the Various
Contrivances by Which British and Foreign Orchids Are Fertilised by Insects."* The book is a sort of biological *Candide*, which,
albeit with the greatest restraint, holds up the very idea of organic design to ridicule and contempt.[18] There are reasons to
think, indeed, that it was written as a deliberate satire on the
Bridgewater Treatises.[19] In a well-known series of letters between Gray and Darwin, the latter affirms that he did have an
ulterior purpose in expounding upon the curious adaptations of
orchids, although it seems a dismal commentary upon the intellects of other workers that Darwin remarks: "No one else has
perceived that my chief interest in my orchid book has been that
it was a 'flank movement' on the enemy." [20] On the other hand,
the purely biological aspects are cast in quite explicit language,
while the metaphysical level is rendered in far more subtle terms.
And the myth seems to have been already current that Darwin
was a mere naturalist and an amasser of facts—a man of limited
intellect and small capacity for abstract thought.

Darwin's work explains how the parts of orchids are so arranged that they facilitate the transfer of pollen by insects. He
demonstrates that such adaptations arose through the modification of some preexisting part, so that a structure which originally

subserved a given function comes to do something quite different and unrelated. He attempted to show, in other words, that structures were not designed with the end in mind of engaging in their present biological role, but rather that they originated as parts adapted to quite different functions. The flower makes use of whatever parts happen to be available, and their availability and utility are purely accidental. Darwin conceived of teleological explanations as presupposing that there has been foresight and deliberate action on the part of some agent to fit the organ to its function. Such an interpretation of teleology might be challenged, and a subtler form of it, such as Aristotle's, might still be supported. But the refutation of the cruder version should remove most of the attraction in final causes, for modern teleology has done nothing but substitute vague, metaphysical influences for the anthropomorphic Diety.

THE SEMANTICS OF DESIGN

Darwin would seem to have recognized that the argument from design is due to verbal confusion. A consideration of the basis of his analysis, although only indirectly related to his scientific work, serves to illustrate the breadth and soundness of his thinking in general. It was essential that he recognize the distinctions between a number of concepts which are readily confused, especially "purpose," "function," and "functional property." For clarity of exposition, we may draw upon a non-biological analogy. Consider two gentlemen, suitors for the same lady's hand, playing a game of "Russian roulette." Player One loads a revolver with a single cartridge, gives the cylinder a spin, cocks the weapon, places the barrel to his head, and pulls the trigger: "click." Player One then hands the revolver to Player Two, who spins the cylinder, cocks the weapon and places it to his head, then pulls the trigger: "BANG!" Player Two has lost, and someone will have quite a mess to clean up. We may observe that the revolver in this episode is involved in an activity, namely a game of Russian roulette. Its role in the game is its *function:* it functions in discharging a cartridge under a particular range of circumstances. We may note, however, that certain actions of the gun have no relation to its role in the game, and are therefore not to be considered among its functions—for example, it may cause the curtains to be splattered with blood. The gun also has a *purpose*, which may be the same as its function, although this need not be the case. The players use it for (and it functions in) blowing their brains out. That is, the players foresee that it will have such an effect and use it with that end in mind. However, it is clear that although the revolver is put to a purpose in the game, the manufacturer in all probability did not design the weapon for that purpose. And it follows that the mere use of a thing as an instrument by no means compels us to believe that it was designed for that purpose or with that end in mind, no matter how well it discharges that office. A pistol may be used, say, as a nutcracker or a paperweight, simply by taking advantage of its fortuitous qualities. The utility of a revolver in the game of Russian roulette depends upon the accidental circumstance that spinning the cylinder allows whether one's brains are blown out to be determined by chance: revolvers were not invented "for" that purpose, although they are well suited to that function.

Darwin showed that biological functions are of an analogous nature: nobody plans anything; opportunities are utilized when they occur, in a sense, at random. And the fact that there is a state of fitness or adaptation tells us nothing per se as to how that condition arose. To understand these matters was an exceedingly significant break with the past. Through a failure to make just such distinctions as those between function and purpose, Aristotle became confused over the entire structure of the physical universe, and insinuated into the structure of Western thought the basic premise that function implies purpose, a notion that poisons our reasoning to this very day.[21] And when it is realized how many contemporary biologists assume that the very existence of an adaptive trait means that it is present because of its selective advantage, it becomes all the more striking how well Darwin grasped the theoretical profundities of his work. The problem with such words as "purpose" and "function" is that they refer to relational properties.[22] One has to supply the right terms for the relation, or else one may make an error. If a thing is to function, it must do so in relation to some action or the like. If it is to have a purpose, it must be given a purpose by somebody. And it may have one function in relation to a given activity, another function in relation to some other activity. Again, it may stand in different purposive relationships to different people. Because "function" and "purpose" are used under rather similar circumstances, they are easily confused. It is readily forgotten that the relation involved in a function has no reference to motivation or design, but simply asserts that something is an integral and important part of some process. Like many biological terms, "function" is such a complicated relation that to formulate what it means is very difficult. The sort of confusion that has arisen exists in spite of the fact that in ordinary discourse biologists have no trouble using it in a meaningful and unambiguous way.[23]

Having defined the terms "function" and "purpose" well enough to make some important distinctions, we are now in a position to take a cold, hard look at the argument from design as it was formulated by Paley and supported by Asa Gray. Paley contends that if one were to come upon a watch, this would necessarily imply that someone, a watchmaker, had constructed it. He argues:

> There cannot be design without a designer; contrivance, without a contriver; order without choice; arrangement, without anything capable of arranging; subserviency and relation to a purpose, without that which could intend a purpose; means suitable to an end, and executing their office in accomplishing that end, without the end ever having been contemplated, or the means accommodated to it. Arrangement, disposition of parts, subserviency of means to an end, relation of instruments to a use, imply the presence of intelligence and mind.[24]

The analogy of a revolver being used for Russian roulette without the inventor or manufacturer foreseeing that it should be so used reduces this argument to an absurdity. Again, one may object that the arrangement of atoms in a crystal implies that there has been some ordering process; but nobody would seriously maintain that this must have been guided by reason. It pays to ask upon just what grounds one may decide that some particular configuration of matter has or has not been the product of intelligent action. What possible reason can one have for saying that the origin of some kind of order must have been such-and-such when one is ignorant of past events and laws relevant to the production of that order? Again, it is by no means justified to identify statements of function with genetic explanations. When someone says that a particular revolver is used in Russian roulette, how much does this explain? It explains absolutely nothing about why that revolver is used in Russian roulette rather than in murder or bank robbery. It does not tell us why it has utility in that office while a rifle does not. The statement of function can fit into all sorts of explanations, to be sure, but only when conjoined with other propositions: a revolver is used in Russian roulette because, unlike an automatic pistol, its discharge may be made to occur at random. Statements of function are descriptions, not explanations, and it is only because they so readily fit into explanatory systems that we read causes into them.[25] It thus seems abundantly clear that the mere fact of its telling us the time, or being so constructed that it can do so, casts no light whatsoever on the origin of a watch.

The real reason for our thinking that a watch implies a watchmaker is provided by considerations other than the mere fact of subserviency to a function. We have sufficient understanding of our everyday surroundings to judge it improbable that any combination of natural forces could produce such an instrument. We know, for example, that if a bunch of wheels and springs are put into a box and the box is shaken, the chances are exceedingly small that a watch will be produced. That is, we reason from known principles, construct a hypothetical example, and calculate (or crudely estimate) the probability of the occurrence. But what of organic order? What is the size of the sample? Where are the premises, the theoretical principles, and the calculations? They are all provided by the imagination. The whole argument from design is an outrageous perversion of the theory of probability, deriving from the notorious "principle of insufficient reason." That is to say, it presupposes that one may, on the basis of personal ignorance, assign an arbitrary probability to the occurrence of an event and hold that probability a priori.[26]

That Darwin was fooled in his youth by the argument from design has parallels in the fact that he did not adequately comprehend the subtleties of probability theory. In later chapters it will be argued that analogous mistakes, involving a comparable assignment of gratuitous probabilities or similar assumptions in the face of ignorance, led Darwin to erroneous views on blending inheritance, on acquired characteristics, and on certain features of chemoreception in plants.[27] His overthrow of the argument from design is also intelligible, for when he used the theory of probability in its proper context, that is, in erecting hypothetical systems of such a structure that the premises have experiential support, no such errors occur. Likewise we may understand why Asa Gray was so much enamored of the argument from design, for he too makes analogous blunders. Consider, for example, the following paragraph from Gray's essay of 1860, *Design Versus Necessity:*

> Now, if the eye as it is, or has become, so convincingly argued design, why not each particular step or part of this result? If the production of a perfect crystalline lens in the eye—you know not how—as much indicated design as did the production of a Dollond achromatic lens—you understand how—then why does not "the swelling out" of a particular portion of the membrane behind the iris—cause you know not how—which, by "correcting the errors of dispersion and making the image somewhat more colorless," enabled the "young animals to see more distinctly than their parents or bretheren," equally indicate design—if not as much as a perfect crystalline, or a Dollond compound lens, yet as much as a common spectacle-glass? [28]

Even though Gray seems partially aware of the premises of the argument, we nonetheless have a most instructive example of the fallacy of division—that is, of the notion that a part must have all the attributes of the whole. If spectacles argue design, then so does glass; if glass, then sand. Again, if Gray's argument were valid, we might carry it to its logical conclusion and maintain that since, in a poker game, we have evidence of a stacked deck when someone receives twenty flushes in a row, then we should accuse a player of cheating if he completes a flush by obtaining a spade at the draw. The situation is analogous to many misconceptions and erroneous criticisms of selection theory needing only a little probability theory and common sense to refute them.

THE TESTING OF FUNCTIONAL AND TELEOLOGICAL HYPOTHESES

Let us now turn to the methodological problem of finding out just what are the purposes or functions of particular attributes and parts. To return to our analogy of the game of

Russian roulette, let us consider how one discovers the function
of an instrument, and how one distinguishes between attributes
generally, uses which were not planned, and design. One answer
is that one might use the same sort of "functional analysis" that
has already been mentioned in relation to Darwin's barnacle
anatomy. As in other scientific work, one would formulate
hypotheses which might explain the properties of the instrument,
predict what should be true if the hypotheses are correct or
erroneous, and look to certain kinds of experience for testing.
Consider, for example, the alternative hypotheses that a given
pistol has been designed as a weapon, or for use in Russian roulette. To subserve its function as a weapon, its parts should be
delicately adjusted so as to insure accuracy: it should be well
balanced, have a straight barrel with lands and grooves, and be
equipped with accurate sights. For Russian roulette, on the other
hand, properties conferring accuracy are superfluous; yet the
cylinder should spin easily, and the pistol should be equipped
with a very short barrel, so that it can be placed against the
head.

Armed with such an analogy, we have a basis for inferring the
more intricate, yet logically comparable, reasoning which Darwin
applied in establishing the functions of particular flowers and in
demonstrating that their properties have not resulted by design.
The approach is particularly obvious in his work on orchids.
Here he gives many examples of how an insect, if it is to pollinate
a plant, must approach the supply of nectar from some particular
direction. It is easy to imagine him reflecting upon how a moth or
a bee should thrust its proboscis into a flower, then considering
how the flower might deposit its pollen on the insect's body.[29]
We find him comparing the unusually long nectary of certain
orchids with the equally remarkable development of the proboscis in certain Lepidoptera.[30] In one instance, he finds no trace
of nectar in an orchid; realizing that this arrangement would be
ineffective in attracting insects, he looks more closely and discovers a source of nectar beneath a membrane. The membrane is
present because the insect must chew through it to get at the
nectar, and the delay provides sufficient time for the pollen mass
to adhere to the insect's body; and it is shown that these particular flowers are peculiar in that it takes a very long time for the
adhesive substance which affixes the pollen to set.[31] It may be
noted that in each of these analyses, the hypothesized mechanism

of pollination, with its subservient adaptive arrangement of parts and attributes, is verified by demonstrating that a wide variety of structures are precisely and intricately adjusted, in a manner explicable in terms of the hypotheses, but not otherwise. It is therefore not the mere fact that single parts are so arranged as to further fertilization that supports the hypothesized function, but the joint implication that all the parts have to be mutually coordinated if the flower is to operate as the hypothesis implies. The arguments form a system.

The investigation of pollination mechanisms depended upon asking the right kinds of questions in the light of the assumption that the male structures must be so placed as to deposit the pollen, and the female structures so arranged as to receive it. Darwin himself explains his approach, going into considerable detail, in his very important work *The Different Forms of Flowers on Plants of the Same Species*. Here he elaborates upon one of his most original contributions: the discovery that certain hermaphroditic plants exist in two forms (fig. 6), in which plants

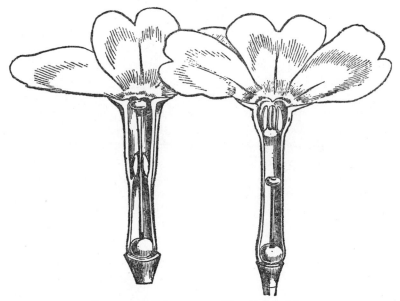

Long-styled form   Short-styled form

Fig. 6 The two forms of *Primula*. The plant is adapted to transfer pollen to, or receive it from, a flower of the other type, but not from one of its own. From *Flowers*.

of one type can cross readily with those of the other, but not with an individual of the same type.[32] This phenomenon was so novel and unexpected that Darwin remarked in his autobiography: "I do not think anything in my scientific life has given me so much satisfaction as making out the meaning of the structure of these plants." [33] The study of heterostyly, as it is now called, is still of great interest, and its mechanisms remain but imperfectly understood.[34] Darwin was led to discover it when he observed that certain plants exist in two forms, one with a short pistil, another with a long pistil. These might be thought mere variations, but this explanation did not seem reasonable, in view of the fact that there were no intermediates. Therefore, some functional explanation seemed in order.[35] Darwin relates how it first occurred to him that he was dealing with an early stage in a change from a hermaphroditic state to one in which the sexes are separate.[36] It is easy to see how his experience with barnacles, in which just such a modification has occurred, might suggest this hypothesis. He reasoned that the long-styled plants of the cowslip, which he observed to have longer pistils and smaller pollen grains, would be incipient females. The short-styled plants, with short pistils and larger pollen grains, would be, as he put it, "more masculine in nature." [37] On the basis of these considerations, he predicted that the long-styled, supposedly more feminine plants should bear more seed. He marked a number of plants from various habitats and weighed the seed from each type of flower. The experiment, first carried out in 1860 and repeated with precautions in 1861, gave results which decisively refuted his hypothesis. By any measure, the long-styled plants were less fertile than the short. Darwin had to seek a better explanation. The second approach is easily understood in the light of his functional analyses of orchids, although he does not explain every step in the investigation. He remarks upon the way a moth's proboscis would fit into the flower and discusses experiments in which he thrust various objects into the opening. It became evident that the pistils and stamens are so positioned that the pollen of one form gets transferred to the pistil of the other, much as the parts are advantageously arranged in flowers with separate sexes.[38] In effect, the heterostyled plants are hermaphrodites which have evolved two mating-types, analogous to the different sexes in more conventional organisms.

Having used functional analysis to develop and support an explanatory hypothesis for heterostyly, Darwin worked out the implications in detail and went on to a series of experimental tests. The basic aim was to see if pollen from a given individual, known to be effective in fertilizing individuals of the other type, was equally successful with plants of its own kind. He carried out four types of crosses. In two of these, which he called "legitimate" crosses, he transferred the pollen as he thought it must be conveyed by an insect, in view of the structure of the flowers: the pollen from a short-styled plant was transferred to a pistil of a long-styled plant, and conversely. In the two other kinds of crosses, which he called "illegitimate," he put pollen from a long-styled plant on the pistil of another long-styled plant, or else that from a short-styled plant on another of the same type. The results clearly showed that in the "illegitimate" crosses, fertility was greatly diminished.[39] There was an obvious disadvantage to crossing in the manner hindered by structure.

In the sequence of reasoning that Darwin employed in his work on heterostyly, we have a classic example of the hypothetico-deductive scientific method, but with something a bit unusual added. In the first episode, there is an observation (namely, heterostyly), a hypothesis which arose from analogy with other cases, a prediction from that hypothesis, and an experimental refutation. In the succeeding parts of the investigation, we have the same phenomena observed, and a comparable development of a hypothesis through analogical reasoning, but there is a further stage of functional analysis. It is clearly the functional analysis which bridges the conceptual gap and makes it possible to come up with a somewhat subtler hypothesis. Yet here, too, the second hypothesis has roots in previous work, and even the decision to use a functional analysis would seem to be based on an analogy. The new ideas had some sort of precursor in earlier experience, and there is no need to invoke great leaps of the imagination in explaining their historical origin. What might be considered a "retroduction" seems far less mysterious when treated as systematic development of various theories, until one of these happens to generate a pattern of relationships which fits in with the phenomena under investigation. Yet we might note, in passing, that the fact of there having been a rigorous hypothetico-deductive test of the theories is not the sole reason why these experiments seem convincing. The so-called H.-D. model does provide a valid explanation of certain basic logical operations in scientific work; but it has its limitations. There is something about the quality of evidence that also needs to be taken into account—and this has largely to do with the variety of phenomena that the hypothesis predicts. It is one thing to be successful in carrying out a variant on a previous experiment, it is quite another to apply a hypothesis in solving a new kind of problem.

Once again, I stress the importance of not confounding the historical derivation of Darwin's ideas—a strictly psychological problem—with the logical problem of deriving their support. In science, just any analogy will not of necessity serve as evidence, and the history of thought is strewn with the corpses of strictly analogical arguments—such as those based on the parallels between macrocosm and microcosm. Analogies in the sense of comparable patterns or logical structures are reasonable places to look for an explanation, and often provide a rich source of important questions. But it would be indeed erroneous to class functional analyses with mere analogies, for they are no mere heuristic maneuvers—they are not simply aids to problem solving.[40] In terms of formal logic, there need be no basic difference between functional analyses and laboratory experiments. They are best grouped with what are called "thought experiments," in which predictions from hypotheses are tested by experience without actually manipulating the objects under study.

HISTORICAL ORIGIN OF FUNCTIONAL PROPERTIES: BARRIERS TO CROSSING

Darwin used his functional analyses to investigate the historical processes which have given rise to present conditions of adaptation. His procedure may be elucidated through an examination of his thinking on the origin of the sterility which prevents illegitimate crossing in heterostylic plants. His reasoning on such matters is of the utmost biological significance, for the evolution of sterility plays an important role in speciation. Were it not for the inability of species to interbreed, they would exchange hereditary material and become increasingly less diverse, as Darwin was well aware. It is therefore advantageous to organisms that they are broken up into separate populations, for these can each be modified and adapted to a particular way of life. Diversity and specialization facilitate the exploitation of different kinds of environments. But how did this sterility come into being in the first place? This question has long been controversial and is still being argued today. But the solution is for the most part a problem not of data but of methodology. Attempts to obtain "direct" evidence have failed to provide an answer but certain hypothetical models have resolved the issue through their empirical implications. It is generally accepted that in most animals sterility arises accidentally, only after populations have been isolated for some time (allopatric speciation), and that geographic separation is usually (perhaps not exclusively) responsible for the isolation.[41] There may be exceptions, and speciation without geographic isolation (sympatric speciation) is considerably more probable in hermaphroditic and parthenogenetic organisms.[42] The reason for considering some kind of geographic isolation necessary becomes evident when one considers a conjectural example. Suppose that there were a species composed of four animals: two males and two females. Should it happen that there were a mutation affecting the genital organs of one of the males, he might not be able to copulate with the females. The deviant male would be sterile, and his genes would be lost to the next generation. If, however, one of the females also changed, she might become altered in such a manner that the two modified forms could reproduce. This might give two species. But how likely is it that both a male and a female not only would change, but would do so in such a manner that the two could reproduce? Unless one invokes foresight or design, it is not at all likely, although it might happen very rarely. Sound biological principles therefore imply that unless there is some special set of circumstances, such as hermaphroditism and facultative self-fertilization combined with polyploidy, speciation without geographic isolation is very improbable.

Mayr has shown that Darwin did not fully realize the degree to which natural selection implies a low probability of sympatric speciation.[43] Because of this oversight, Darwin de-emphasized the importance of geographical isolation. Once one understands the issues, the problem seems quite straightforward; but the number of biologists who have missed the point shows how easy it is to go astray in such matters. The difficulties are even more obvious when we see that although Buffon, in the eighteenth century, evidently thought of the model which elucidates such matters, he used this same argument to argue against the formation of new species—he overlooked the role of geographical isolation.[44] The solution to this kind of problem cannot be provided simply by looking at individual situations which seem to imply that speciation has occurred through one or another mechanism. One can hardly expect the data to be so copious as to tell us that there was no period in which the populations in question were not isolated. In general, arguments for sympatric speciation, especially in animals with the usual type of sexuality, are based upon the theorist's inability to imagine a geographical isolation mechanism—a truly pretentious hypostasizing of ignorance. The sound way to evaluate the conflicting theories of sympatric and allopatric speciation is to elaborate a hypothetical model, derive its implications, and look to the facts for corroboration. The argument must be a statistical one, for no individual pattern of properties will do, just as an early death of one policy-holder is no reason for an insurance company to throw away a set of mortality tables. Confirmation of the necessity for geographic isolation has for the most part come from distributional studies. The theory implies, for example, that (1) the amount of speciation will correlate with the presence of geographical barriers to distribution; (2) the groups that speciate extensively in a given habitat will be those having difficulties of dispersal in that habitat; (3) the forms which speciate extensively in areas where there are few geographic barriers will be those with special types of reproductive relationships which account for their being exceptions. These predictions are all borne out by the available data.[45] It is of no slight interest that this kind of argument resembles the ones Darwin used to test his coral reef hypothesis, and in his geographical verification of evolution. For strikingly similar reasons, it compels assent. Likewise, those who disagree with the conclusions do so because they do not understand the argument; the possibility of exceptions has never been an issue.

Perhaps one reason why Darwin de-emphasized the importance of isolation in speciation was that he tended to conceive of species as typological classes rather than as populations. It is only when one constructs a model of the population interactions, that the breakdown, through migration, of incipient barriers to crossing, is easily imagined. But Darwin had another kind of objection to the selective origin of sterility. His argument derives from a different kind of theoretical perspective, one which is more typical of his approach, since it involves the effects of differences between individuals. His model predicts that sterility will not arise by selection, but it has no bearing on the issues of allopatric versus sympatric speciation. On this basis one may corroborate this particular reconstruction of Darwin's thinking, for it not only explains both discoveries and oversights, but it also shows that his approach had analogous effects on his treatment of other problems.

Much of Darwin's work on sterility appears in his book *The Effects of Cross and Self Fertilisation in the Vegetable Kingdom.* He relates that he was led to undertake the eleven years of experimentation treated in this work when he unexpectedly observed that self-fertilization in plants results in very low fertility.[46] The fact that he did not expect such pronounced effects argues for the idea that he had no theoretical reason to expect a physiological disadvantage for inbreeding. Rather, he evidently was looking for an advantage to crossing, or else a disadvantage to selfing, that would appear after several generations. His experiments ultimately demonstrated that plants vary greatly in their ability to fertilize themselves: Some few plants are clearly adapted to self-fertilization, but a major portion have most intricate mechanisms insuring cross-fertilization.

Darwin recognized that self-sterility would not be favored by natural selection, in spite of any long-term advantages to the species. As he put it: "The means for favouring cross-fertilisation must have been acquired before those which prevent self-fertilisation; as it would manifestly be injurious to a plant that its stigma should fail to receive its own pollen, unless it had already become well adapted for receiving pollen from another individual." [47] Evidently this means that the advantages of fertility, rather than of sterility, must have been the selection pressure that produced the block to self-fertilization. This assumption would not contradict the possibility that, once cross-fertilization had been assured, the accentuation of self-sterility might be favored.[48] The issues involved in this problem are rather obscure, but they may be clarified by reference to the point already discussed, of what Darwin considered an individual to be. We may recall how, in *The Origin of Species*, Darwin pointed out that selective advantage in the case of a hive of bees is determined by the reproductive success of the hive as a unit. Darwin conceived of selection as only occurring through advantage to individuals (including social units). Thus, one may explain why individual bees are sacrificed to the advantage of the group as a whole when they kill themselves in stinging an intruder in the hive. In like manner, the altruistic behavior of man is clearly intelligible in terms of the advantage it confers upon the family. But any such relationship is not evident with respect to plants. There is no obvious mechanism through which an altruistic self-sterility could be favored in these nonsocial organisms. Yet perhaps workable mechanisms might be postulated if one assumes that there is some kind of social interaction between the parent and progeny, or between progeny that are produced through selfing and crossing in the same parent. Such interactions are very difficult to deal with, as may be seen from the example of brood size in birds.[49] It is known that birds may be able to lay more eggs than they actually deposit in the nest. One might think that selection would favor the laying of as many as is physiologically possible, but such is not the case. The parents are able to feed only a limited number of offspring, and if there are too many young, the whole brood perishes. Hence, a bird which lays fewer eggs may raise more young because its energies are used more effectively. Many analogous cases are far more complicated and require sophisticated mathematical techniques for effective analysis.[50] Perhaps Darwin oversimplified the problem. He certainly approached it in the proper manner, and there is no doubt that he understood the logical issues. Darwin and Wallace exchanged a series of letters in which the whole question is clearly formulated.[51] Wallace went to extreme lengths to construct a model of selection for sterility, but Darwin was unconvinced by these efforts, and rightly so, for they were exceedingly farfetched. The dispute is particularly revealing, for it casts much light on their respective attitudes toward methodology and metaphysics. Wallace, who was always seeking plausible reasons, held, erroneously, that the inability to account for sterility argues against natural selection.[52] Darwin, who, by contrast, was out to develop and test his theory, held that it was by no means essential that natural selection explain the origin of every adaptive trait. One should expect

an occasional biologically advantageous feature to arise by accident, much as revolvers have proved useful in Russian roulette.
The same kind of disagreement, with Wallace and with others,
occurs in various contexts, clearly demonstrating Darwin's superiority as a theoretician over his contemporaries. It was because
of Darwin's innate capacity for logic that he had to remark to
Wallace: "We shall, I greatly fear, never agree." [53]

Darwin seems to have concluded that one may explain the
origin of sterility as an accidental consequence of changes which
favor outbreeding. As this explanation seemed to him sufficient,
he did not pursue the matter any further. The accidental origin
of sterility is analogous to a number of other notions which Darwin entertained. His embryological models implied that certain
properties of a neutral or even deleterious nature would arise as
side effects of selection for other properties (pleiotropy). In
chapter 9 it will be shown that Darwin, in his treatment of avian
coloration, overlooked a number of relationships not only because his hypothesis failed to predict them, but also because the
data as a whole were explicable through pleiotropy and other
factors.

It would give an altogether misleading impression to represent
Darwin's inference that sterility arose as a pleiotropic effect as if
it were derived solely from theoretical considerations. His hypothesis had many empirical implications, and observation bore
out his theoretical expectations. By way of illustration, we may
consider another example from his work on heterostylic plants.
One species, *Lythrum salicaria*, is remarkable in being heterostylic, but instead of the usual two kinds of flowers, it has three
(fig. 7).[54] It is a hermaphrodite, but it has three different mating
types, the presence of which is analogous to its having three
different sexes. The pistils and stamens are so positioned that
pollen is deposited and received at three different spots on an
insect's body. One kind of flower has a very short pistil and two
groups of stamens: one very long, and the other of intermediate
length. Another form has a very long pistil, and the sets of
stamens are very short, or of moderate length. In the third, the
pistil is of intermediate length, while the stamens are very long
and very short. Darwin undertook a series of experiments similar
to those already described, in which he compared the effects of
legitimate and illegitimate crosses. He tried the various possible

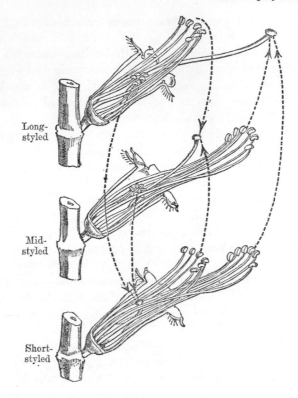

Long-
styled

Mid-
styled

Short-
styled

Fig. 7 Heterostyly with three forms. Arrows indicate "legitimate" crosses. From *Flowers.*

unions between the different types and obtained quantitative data on their relative fertility. The degree of sterility depends, he found, on the distance between the pistil and the stamen. In, say, a cross involving a very short pistil and a very long stamen, the sterility is greater than in a cross involving a pistil of intermediate length and either a very short or a very long stamen. The sterility does insure that crosses will be legitimate. However, the most advantageous arrangement, from the point of view of preventing illegitimate crosses, would be one which is most efficacious not in blocking crosses between very short and very long stamens and pistils, but rather in blocking crosses between long and short sexual parts and those of intermediate length. That is to say, the greatest need for sterility is where there is the least distance between the stamen and the pistil, but actually the closer parts have less of a barrier to crossing than do more distant ones. For this reason Darwin argues: "We are led, therefore, to conclude that the rule of increased sterility in accordance with increased inequality in length between the pistils and stamens, is a purpose-less result, incidental on those changes through which the species has passed in acquiring certain characters fitted to ensure the legitimate fertilisation of the three forms." [55] One might argue that this inference goes a bit too far beyond the data. Nonetheless, considered in its proper context, the argument is valid.

THE REFUTATION OF DESIGN

Having demonstrated how hypotheses concerning the mechanism of origin of particular attributes may be tested by functional analysis, we are in a position to undertake a more detailed treatment of Darwin's arguments against teleology. He was able, as we have seen, to show reasons for thinking that properties of organisms have arisen, in a sense, by accident. Sterility was not planned, or even produced, through its selective advantage, but arose through pleiotropy. Most important, organic mechanisms may be shown, through analogous considerations, to have been haphazardly thrown together, out of whatever materials the moment happened to supply. In his book on orchids Darwin argues that new structures are invariably elaborated out of preexisting components which, by coincidence, already happen to possess useful physical or chemical properties. Thus he remarks upon how the glue which originally held pollen to the stigma is used, in modified form, to affix pollen masses to the bodies of insects.[56] Likewise, he points out that many properties, readily explained as vestiges of an earlier function, make no sense as features subservient to some end. Therefore he argues:

> Thus every detail of structure which characterizes the male pollen-mass is represented in the female plant in a useless condition. Such cases are familiar to every natural-ist, but can never be observed without renewed interest. At a period not too far distant, naturalists will hear with surprise, perhaps with derision, that grave and learned men formerly maintained that such useless organs were not remnants retained by inheritance, but were specially

created and arranged in their proper places like dishes
on a table (this is the simile of a distinguished botanist)
by an Omnipotent hand "to complete the scheme of
nature." [57]

From the analysis given earlier in the present chapter, where it
was shown that a thing may have a different purpose or function
in relation to different activities, it becomes readily apparent
what Darwin means when he says:

Although an organ may not have been originally formed
for some special purpose, if it now serves for this end, we
are justified in saying that it is specially adapted for it. On
the same principle, if a man were to make a machine for
some special purpose, but were to use old wheels, springs,
and pulleys, only slightly altered, the whole machine,
with all its parts, might be said to be specially contrived
for its present purpose. Thus throughout nature almost
every part of each living being has probably served, in a
slightly modified condition, for diverse purposes, and has
acted in the living machinery of many ancient and distinct specific forms.[58]

In other words, the same part may stand in different kinds of
adaptive relations, and to overlook the fact that the relations are
not the same leads to verbal confusion—without which there can
be no argument from design. Once more we see that Darwin,
who has been said to have lacked philosophical sophistication, was
not without a certain competence in linguistic analysis.

But suppose that the flowers had been designed by an intelligent Artificer—would He really have constructed orchids according to the observed patterns? Perhaps He could have come up
with something better. Darwin stresses the significance of the
remarkable "prodigality of resources,—for gaining the very same
and, namely, the fertilisation of one flower by pollen from another plant." [59] From the theory of natural selection it follows
that different variations, occurring fortuitously in different lineages, should come to produce alternative adaptive configurations
having the same effect. But if the Divine Intellect has designed
organisms so that each is an effective mechanism for dealing with
pollen transfer, why are there so many designs? Why did He not
endow all species with the single, most perfect instrument? Elsewhere Darwin elaborates upon this same basic argument, when he

points out that Helmholtz had shown how the human eye, a stock example of the perfect fitting of organs to their functions, is really a most inefficient contraption.[60] Darwin had stood the argument from design on its head, and drawn precisely opposite conclusions out of the same kind of data.

Asa Gray, while never denying the importance of natural selection, attempted to save the argument from design by the curious stratagem which has already been mentioned. He held that perhaps the Omnipotent Creator had foreseen and planned the development of life through natural selection. Thus, natural selection would be the instrument of God in attaining his conscious and deliberate ends. Each event would thus be part of a preordained plan. Gray gives the following argument:

> Wherefore, so long as gradatory, orderly, and adapted forms in Nature argue design, and at least while the physical cause of variation is utterly unknown and mysterious, we should advise Mr. Darwin to assume, in the philosophy of his hypothesis, that variation has been led along certain beneficial lines. Streams flowing over a sloping plane by gravitation (here the concept of natural selection) may have worn their actual channels as they flowed; yet their particular courses may have been assigned; and where we see them forming definite and useful lines of irrigation, after a manner unaccountable on the laws of gravitation and dynamics, we should believe that the distribution was designed.[61]

Gray's idea bears close resemblance to the preestablished harmony of Leibniz, and to the embryological notions of preformation and *emboîtement*. His argument, as it was presented, could be treated as a scientific hypothesis, for it had empirical implications. Darwin, in his *Variation of Animals and Plants under Domestication*, accepted the challenge; he worked out the consequences of the hypothesis and refuted it. For he saw that the occurrence of evolution through natural selection results from the interaction of all the inhabitants of the earth, and that these interactions would likewise have to be preordained, down to the last symbiosis between bee and flower. So also would the distribution of every organism that had ever lived, and all the events of geological history. Every single variation would have to be foreseen and planned. Worse still, for Gray's argument, the Creator would have had to plan not only the beneficial variations, but the harmful ones as well. As a consequence of his embryological reasoning, Darwin realized that variations are not simply beneficial or otherwise. Pleiotropy implied that advantages would be purchased at the price of defects. God must have foreseen both good and ill, and yet it was evident that he could have chosen better: the Divine Artificer had, for instance, constructed some female fowl, but not others, with spurs that interfere with the incubation of the eggs.

Darwin's statements on design on the installment plan, following as they do a widely condemned theory of heredity, are rarely noted.[62] It is therefore overlooked how clever is his argument, which displays no small power of analogical reasoning. The formation of adaptive structures through fortuitous variation is compared to the construction of a building using uncut stone selected from a pile at the foot of a precipice. He disposes of the notion that his theory need explain variation, on the grounds that to understand the work of the architect, one need not inquire into the forces that gave each stone its particular shape. Then, with his usual subdued, yet pointed irony, he concludes the work with a devastating rejoinder to Gray:

> The shape of the fragments of stone at the base of our precipice may be called accidental, but this is not strictly correct; for the shape of each depends on a long sequence of events, all obeying natural laws; on the nature of the rock, on the lines of deposition or cleavage, on the form of the mountain, which depends on its upheaval and subsequent denudation, and lastly on the storm or earthquake which throws down the fragments. But in regard to the use to which the fragments may be put, their shape may be strictly said to be accidental. And here we are led to face a great difficulty, in alluding to which I am aware that I am travelling beyond my proper province. An omniscient Creator must have foreseen every consequence which results from the laws imposed by Him. But can it be reasonably maintained that the Creator intentionally ordered, if we use the words in any ordinary sense, that certain fragments of rock should assume certain shapes so that the builder might erect his edifice? If the various laws which have determined the shape of each fragment were not predetermined for the builder's sake, can it be

maintained with any greater probability that He specially ordained for the sake of the breeder each of the innumerable variations in our domestic animals and plants;—many of these variations being of no service to man, and not beneficial, far more often injurious, to the creatures themselves? Did He ordain that the crop and tail-feathers of the pigeon should vary in order that the fancier might make his grotesque pouter and fantail breeds? Did he cause the frame and mental qualities of the dog to vary in order that a breed might be formed of indomitable ferocity, with jaws fitted to pin down the bull for man's brutal sport? But if we give up the principle in one case, —if we do not admit that the variations of the primeval dog were intentionally guided in order that the greyhound, for instance, that perfect image of symmetry and vigour, might be formed,—no shadow of reason can be assigned for the belief that variations, alike in nature and the result of the same general laws, which have been the groundwork through natural selection of the formation of the most perfectly adapted animals in the world, man included, were intentionally and specially guided. However much we may wish it, we can hardly follow Professor Asa Gray in his belief "that variation has been led along certain beneficial lines," like a stream "along definite and useful lines of irrigation." If we assume that each particular variation was from the beginning of all time preordained, then that plasticity of organisation, which leads to many injurious deviations of structure, as well as the redundant power of reproduction which inevitably leads to a struggle for existence, and, as a consequence, to the natural selection or survival of the fittest, must appear to us superfluous laws of nature. On the other hand, an omnipotent and omniscient Creator ordains everything and foresees everything. Thus we are brought face to face with a difficulty as insoluble as is that of free will and predestination.[63]

Once again, Darwin has dealt, and dealt successfully, with a problem in analytical philosophy. Yet one may observe that in a sense he accomplished something more. He took the problem of design out of the realm of metaphysics and made it a scientific issue. He asked the question of design in the form of a scientific hypothesis, inquiring just what sort of empirical consequences one would expect *if* God had structured each organic being according to the same kinds of standards that an engineer or craftsman might demand. The facts dô not bear out the predictions, and insofar as it has a scientific basis, the argument from design is refuted. The properties of organisms are all readily explained as the consequence of piecemeal, blind, and often, in the long run, deleterious changes. The course of evolution shows no more evidence of foresight than does a river's flow to the sea.

With the publication of *The Origin of Species*, natural theology was doomed to extinction. Naturally the metaphysician who desires to read some kind of plan or design into nature may argue that inefficiency does not contradict efficiency. It is like the problem of evil: somehow, suffering is necessary. But such arguments can serve to establish anything, and as Darwin showed, if one accepts them, there are no grounds for rejecting arguments which lead to the opposite conclusion. Although in some quarters vitalism still gives off an occasional dying gasp, educated people no longer look to science to justify their religious beliefs. Perhaps, by default, the real victor will be, not science, but that religion which derives its support from mystical experience rather than philosophical argument.

No doubt many will be inclined to argue that the foregoing discussion is superficial—that a number of profound philosophical issues have been left unexamined. I answer, on the one hand, that philosophical issues remain, but that these have nothing to do with the relevant metaphysical issue of whether the structure of organisms manifests design. On the other hand, an answer to the question of why people continue to interpret biology in teleological terms has little to do with philosophy. It is a psychological problem, comparable to an optical illusion and fully explicable in terms of natural science. We need only invoke the hypothesis of Sapir and Whorf, which asserts that the structure of our language determines the perception of reality.[64] When we habitually say that "a bird's wing is for flying," our thought is automatically channeled into the contexts in which the word "for" ordinarily occurs. There is no difficulty in constructing an artificial language—or more accurately speaking a modified natural one--more suited to the unbiased apperception of the universe. The practicing scientist learns to structure his thought upon a more abstractive, theoretical level, and uses the natural language only for communication. At an early stage in his training, he must translate consciously from one language to another, often using intricate circumlocutions. When the biologist learns this art, the problem of teleology simply vanishes; but not everyone has been blessed with this sort of discipline. It is time for philosophers to realize that biologists seek to convey propositions, and not any metaphysical nonsense that might be read into a sequence of words. The issue is not whether there is purpose in nature, but whether language will be our slave or our master.

What has been said here contradicts the widespread opinion that Darwin was not interested in philosophical ideas. For it has been shown that he concerned himself, successfully, with a number of difficult logical and metaphysical problems. It is true that he restricted the philosophical pronouncements in his writings to problems closely related to his scientific work. This, however, may be taken as a sign of wisdom rather than disinterest. It is abundantly clear that Darwin rejected questions of ultimate reality as unanswerable. His ability to demonstrate that certain metaphysical notions have roots in verbal confusion shows that he derived this view on purely rational grounds. His position on metaphysics scarcely differs from that of a number of modern philosophers, and we can hardly blame him for being a century ahead of his times in yet another field. Those who condemn Darwin as incompetent in philosophy do so either from ignorance of his ideas, or because they, personally, would prefer to reject his conclusions.

7. *Variation*

Variation is a fundamental concept in Darwin's evolutionary theory. Natural selection can not operate unless organisms differ from each other. Yet, as has already been pointed out, it was in no way essential to explain how, or why, organisms vary. The cause of variation was not relevant to his argument; if there was variation, and if its effects were not somehow counteracted, natural selection must necessarily bring about evolutionary change. On the other hand, there were reasons to wonder if some unknown forces or circumstances do not render selection impotent. To be sure, Darwin was able to draw upon artificial selection for examples of how inherited differences are, in fact, selected, and how change does result. Yet the conditions under which artificial selection occurs are not quite the same as those in nature. It was suggested that when two organisms are mated, their offspring ought to be intermediate in character between the two parents. But any new difference, however advantageous it might be, would become, under natural conditions, progressively less pronounced with each generation, as the variant offspring mated with nonvariant forms. The animal breeder, it was thought, overcame this effect by keeping the extreme variants totally isolated from the original population. After the rediscovery of Mendel's laws (*ca.* 1900), it became clear that this objection was in error. A characteristic is not inherited as a compromise between the parental attributes. Organisms receive from their parents discrete functional units which control development. The functional units ("genes") maintain their integrity as they are passed from one generation to another. Therefore, they cannot be swamped out by the presence of other genes. However, the relationship between hereditary material and inherited trait is no simple matter of one-to-one correspondence. Indeed, the term "gene" is in many ways much too abstract, as it does not designate the sort of single, discrete entity that was originally hypothesized. The presence of visible properties depends upon the particular combination of genes that an organism possesses, conditioned, however, by environmental factors. Since it may happen that many genes control the structure of a given organ, and since the environmental response varies considerably, matings of different organisms may result in intermediate progeny with all sorts of gradations between them. From this and other causes, one tends to get the impression that the progeny are a blend between the character of the two parents. It is, therefore, easy to understand why a "blending" theory of inheritance, rather than a "particulate" one such as Mendel's, was for the most part presupposed by Darwin and his contemporaries.

If we inquire into the underlying causes of the failure to elaborate a "particulate" theory, answers are not far to seek. The difficulties arose partly because of the intrinsic complexities of the problem, but perhaps more through an inability to formulate the proper questions. One hindrance was a continuing tendency to reason typologically when dealing with hereditary phenomena. Even Darwin never fully escaped from the habit of thinking in terms of idealized classes. It was entirely natural to conceive of two idealized animals, each with its own "character," mating and producing offspring of intermediate "character." A distinct break with the past was necessary before biologists would, as a matter of course, think of individuals, each with a unique set of characters, coming together to give rise to a progeny consisting of individuals, each in turn unique in its particular combination of characters. And the difficulty was compounded by a failure to distinguish between the inherited hereditary material (genotype) and the visible properties (phenotype).[1] The data were such a confused jumble of phenomena resulting from the interaction of so many different causal influences, that any attempt to obtain a solution through the less sophisticated forms of induction was bound to fail. An answer would not be found until an appropriate hypothetical model was erected on the basis of the proper theoretical considerations, and until a means was devised for separating the effects of the proposed mechanism from the welter of raw data. There is every reason to think that Mendel proceeded

by a modern, hypothetico-deductive method, beginning with
speculation, and did not simply obtain his hypothesis as an automatic product of breeding experiments.[2] Since Mendel's approach required considerable restructuring of our way of thinking about the problem, it is easy to understand why his work
went unnoticed for some thirty-five years, until the development of cytology and other sciences, coupled with the persistent
failure of "blending" theories, directed the attention of other
workers along the same lines.[3]

SAVING THE HYPOTHESIS

Whatever may have been Darwin's reasons for adopting
an erroneous view of heredity, it is clear that it posed a difficulty
for his theory. As a conscientious scientist, it was incumbent on
him to answer the objections. He might reply, and to some extent
he did, that he could support his theory by its empirical predictions, and on this basis might argue that the purely theoretical
considerations were somehow misleading. Or he could maintain
that since so little was known of heredity, it could hardly serve as
a rigorous basis for argument. Alternatively, he might save his
theory by showing that under certain conditions selection could
operate in spite of blending; it is abundantly clear that this was
his major line of defense. Finally, he might abandon natural selection and invoke other forces as the causes of evolution. One
writer after another, evidently following R. A. Fisher, has stated
that Darwin adopted this last policy.[4] The alleged fact of Darwin's having elected to de-emphasize natural selection in order to
win the battle for evolution, has been one of the major grounds
for recent attacks on his intellect and character.[5] But however
true it may be that Darwin invoked other causes as explanatory
factors in evolution, this does not of itself demonstrate that he
did so to overcome the problems brought about by blending
inheritance, or that he really gave up natural selection in the face
of criticisms. On the contrary, the facts seem adequate to demonstrate that although Darwin considered blending a threat to his
theory, he never concluded that it posed insuperable difficulties
for his hypothesis. It may also be shown that those alterations in
his theory which accounted for blending were minor, that what
his critics call "Lamarckian" ideas were in no way invoked to

save his theory, and that the hypothesis of pangenesis—which has aroused most resentment of all—was devised to account for purely empirical observations.

It is fairly well recognized that Darwin gave a partial answer to the problem of blending by invoking isolation.[6] The smaller the breeding community, the less the degree to which the properties of a single, deviant individual would be diluted, so to speak, by repeated crossing with nonvariant forms. A small population of organisms could develop a characteristic and expand to a sufficient number that its properties would not be swamped out by blending. It was clear by analogy with domestic animals that a variant form could be perpetuated under such conditions. (We may here recall the "founder principle" and see that there is some truth in such thinking.) However, the objection could always be raised that perhaps the conditions under which artificial selection takes place are not strictly paralleled by those in nature; and it is clear that Darwin concluded, at least with respect to the isolation of single, variant forms, that the necessary isolation does not, in fact, occur. One might think that Darwin became disenchanted with isolation because small populations would lower the probability that a useful variation would occur.[7] But it is clear from his statements that his real reason was that artificial selection has proceeded by quite different mechanisms. He observes that "a new race can readily be formed by carefully matching the varying offspring from a single pair, or even from a single individual possessing some new character; but most of our races have been formed, not intentionally from a selected pair, but unconsciously by the preservation of many individuals which have varied, however slightly, in some useful or desired manner." [8] Likewise in later editions of *The Origin of Species,* he rejects, by similar comparisons, the notion that large changes in structure have been responsible for evolutionary change.[9] On the basis of empirical evidence, therefore, Darwin rejected isolation and fell back on a variety of other explanations.

Largely in order to make certain that the theory of natural selection was consistent with the actual mechanisms of change, but also to overcome such theoretical difficulties as blending inheritance, Darwin undertook a long series of investigations into the nature of variation and heredity. These studies embraced both wild and domesticated plants and animals. His ideas are briefly summarized in *The Origin of Species* and *The Descent of Man*, and are developed at great length in *The Variation of Animals and Plants under Domestication*. A work on variability in a state of nature was planned but never written.[10] These works incorporate what Darwin evidently felt was an adequate solution to the problems raised by blending inheritance. However, partly owing to some erroneous features in his theory of inheritance, and partly because of the complexity of his thinking, his insights have been most imperfectly understood. To interpret Darwin's answers, we must first analyze his questions; however, these were never expounded in explicit terms. The discussion of such matters is largely organized around concrete examples, and the threads of the argument extend throughout the approximately nine hundred pages of *The Variation of Animals and Plants under Domestication*, and even into *The Descent of Man*. Darwin's investigation appears to have been concerned with the following topics: (1) the extent of variation; (2) the conditions under which organisms vary; (3) the laws of variation; (4) the relationships between variation and inheritance; (5) the interrelationships between variation, inheritance, and selection under natural and artificial conditions. In relating these topics to evolution, he had to demonstrate that his various theoretical ideas could provide a consistent explanation for the facts of variation, and in addition that these facts could not be otherwise explained.

VARIATION AND EMBRYOLOGY

Basic to any interpretation of Darwin's work is an understanding of how he used the word "variation." To taxonomists, "variation" often means the differences between the individuals under study. To Darwin, "variation" meant the process which generates these differences. This is no minor distinction, for his shifting of the emphasis from product and group to process and individual is a major reason why Darwin succeeded to the extent that he did. He conceived of variation as a problem, not of genetics, but of embryology. Recognizing that the laws of heredity were so ill-understood as to be almost useless in theoretical application, he shifted to the developmental aspects of the problem. Like a military strategist obtaining victory by choosing the right battlefield, he elaborated this theory along lines which would

eliminate heredity from relevance to the questions at issue. Because he concerned himself less with the laws of heredity transmission of the genotype than with those which govern the expression of the phenotype, he was able to overcome many of the
difficulties that result from not distinguishing between the two.

A major conclusion derived from Darwin's embryological
analysis of variation was the discovery, one of his most significant
contributions to knowledge, that variation is not random but
fortuitous: it occurs according to its own laws, and these are
basically those of developmental mechanics. Much as in human
culture, where the availability of one or another kind of building
material influences the kind of architecture that will prevail
(wood in Japan, brick in Mesopotamia, stone at Rome), the laws
of variation have profound effects upon the course of evolution.
The contingencies of developmental mechanisms vastly complicate the problems of evolutionary theory. But Darwin foresaw
the implications of such phenomena, realizing that an understanding of the laws of variation would provide invaluable explanations for the facts of evolution, and that it could serve as a
guide to the further elaboration of his theory.

Although Darwin recognized that he could treat variation as a
brute fact and simply ignore the problem of its cause, he did
propose one explanation for it. This was the deceptively simple
view that organisms vary when their conditions of existence
change. One might interpret this to mean that he had confused
inherited changes with the noninherited responses of individuals
to environmental stimuli. Thus, one might think he meant that
such responses as the darkening of human skin in summer, or the
growth of a dog's fur with the onset of cold weather, would be
the source of variations which, in his way of thinking, are elaborated by selection. There may be some truth in this interpretation
of Darwin's views, but it is nonetheless misleading, for it draws
no distinction between adaptive and nonadaptive environmental
effects. The significance of this distinction may be seen where
Darwin surveys the effects of changes in climate among domesticated animals and finds that although the wool of sheep may
degenerate in warm climates, artificial selection may counteract
the change. Hence, any effect of the environment must not be
thought of as giving direction to the changes that take place. He
concludes: "In all such cases, however, it may be that a change of

any kind in the conditions of life causes variability and subsequent loss of character, and not that certain conditions are necessary for the development of certain characters." [11] In other words, the relationship may be quite random between the properties which develop after a change in conditions, and the nature of the conditions themselves. The type of variation which occurs depends, not upon ecology, but upon embryology. There is a certain residuum of truth in the notion of new conditions evoking a variation which can be a basis for evolution through natural selection. The expression of a gene in the phenotype depends on the conditions under which development happens to take place. And a particular trait may appear in some environments but not others—as in the tanning of human skin. Since a gene must be expressed before selection can act, a shift in habitat may easily bring about a change in the properties that are subjected to selection. Darwin supported his general ideas about such matters by pointing out such phenomena as the fact that the amount or kind of change in circumstances does not correlate with the amount or kind of variation. He used one of his favorite techniques, prediction of distributional consequences, to show that the nature of the changes which develop with new conditions of life does not determine the course of evolution. He argues, for example: "Hardly any fact shews us more clearly how subordinate in importance is the direct action of the conditions of life, in comparison with the accumulation through selection of indefinite variations, than the surprising differences between the sexes of birds; for both will have consumed the same food, and have been exposed to the same climate." [12] Darwin was never able to establish, with any certainty, the degree to which the environment acts directively in evoking particular types of variations. But the following would seem to be his definitive view: "We are thus driven to conclude that in most cases the conditions of life play a subordinate part in causing any particular modification; like that which a spark plays, when a mass of combustibles bursts into flame—the nature of the flame depending on the combustible matter, and not on the spark." [13]

The most fundamental of Darwin's ideas on the relationships between development and evolution is perhaps that of "correlated variation." This concept, which pervades his works, may, as has already been noted, be identified with the pleiotropy of modern biology. The idea arose very early in Darwin's thinking. In his fourth notebook on transmutation of species we read: "Thinking of effects of my theory, laws probably will be discovered of correlation of parts, from the laws of variation of one part affecting another." [14] De Beer remarks, with some appropriateness, that overemphasis on this kind of relationship predisposed Darwin to overlook particulate inheritance.[15] But one could argue just as well, if not better, that those who came after him underemphasized it. The classical theory of the gene, with its corpuscular functional units, is now seen to have been an oversimplification. It gave too much attention to parts and too little to the relations between them. Our newer understanding of position effects, allometry, and pleiotropy makes Darwin's thinking seem almost prophetic. It is particularly instructive to consider the degree to which his major evolutionary works are organized around the theoretical development of just such ideas, for they manifest the kind of attitude toward relational properties that modern biology has only begun to develop. And he fully understood what he was doing, as may be seen when he says: "The wattle in the English Carrier pigeon, and the crop in the Pouter, are more highly developed in the male than in the female; and although these characters have been gained through long-continued selection by man, the slight differences between the sexes are wholly due to the form of inheritance which has prevailed; for they have arisen, not from, but rather in opposition to, the wish of the breeder." [16] Clearly he has answered, through strictly empirical evidence, the objection which naturally comes to mind, that such a pleiotropic effect would be counteracted by selection for what we now call modifier genes. Again, *if* such variations occur, *then* the pleiotropic effects do not take place; but nothing tells us what variations to expect.

CONCEPTUAL DIFFICULTIES

That Darwin did not advocate a simpleminded version of the "blending" theory may readily be substantiated by reference to his relational point of view. He takes pains to enumerate many characters which do not blend, such as the properties of the "turnspit" dog, which has the same kind of short-leggedness that may be seen in human achondroplastic dwarves.[17] The problem

of deciding what were Darwin's actual views on the importance
of blending is complicated by the fact that he does not always
distinguish between the blending of populations and the blending
of properties in a new individual. Darwin's failure to make the
proper distinctions in this and analogous contexts had a number
of effects on his thinking. We have seen how deleterious was his
tendency to think in terms of idealized classes. He did not rigorously distinguish the properties of the individual from those of
the population, nor the differences which characterize the species
and the species itself. And he makes little distinction between
"character" and "characters." [18]

One may establish the particular kind of fallacious reasoning
that led Darwin to invoke blending inheritance by showing that
he makes the same kind of mistakes in analogous situations. Such
a parallel error occurs in his work on insectivorous plants.[19]
Darwin undertook a long series of experiments on the sensitivity
of *Drosera* to a variety of chemicals, with the end in mind of
comparing the physiology of chemoreception in plants and animals. He was astonished at the minute concentrations which
sufficed to elicit a response, and even repeated his experiments
with additional controls. A comparison of the sense of smell in
animals, however, convinced him that the quantity was by no
means out of the ordinary. Yet he expresses himself in a manner
which at first seems odd, when he says that the particles detected
by animals "must be infinitely smaller." [20] Why smaller *particles*,
when dilution would explain the facts? It seems that Darwin must
have reasoned from an analogy with the breakdown of macro-
scopically visible objects and extended it indefinitely. The division and blending of the phenotype is obviously comparable. In
both instances, a particular kind of explanation was attributed to
a phenomenon without giving any support for the basic premise.
There is no empirical foundation for the idea that traits blend
indefinitely, or that dilution proceeds exclusively by division.
Deriving such important inferences from untested theoretical
assumptions was clearly a defect in Darwin's methodology,
which demonstrably led to serious mistakes. When, on the other
hand, the basic premises are treated as hypothesis—as in the coral
reef theory and that of natural selection—the possibility of such
errors is far less of a problem, for they result in erroneous predictions. In the previous chapter, the assignment of gratuitous

probabilities in the face of ignorance has already been condemned with respect to the argument from design. Yet before we judge Darwin too harshly, it seems wise to reflect on the frequency of such mistakes in modern biology. A good example may be drawn from the very prevalent assumption by systematists that the "characters" of taxonomy are, like the component nucleotides of a chromosome, limited in number. Thus, an eminent modern taxonomist writes: "No one can say for certain, but it seems probable that the Tasmanian 'wolf' and the placental wolf have more characters in common than the Tasmanian 'wolf' and a kangaroo." [21] The problem is still with us. Indeed, we may expect it to endure as long as biologists remain indifferent to the philosophical aspects of their subject.

THE MECHANISMS AND LAWS OF VARIATION

Darwin's approach to variation obviated the difficulties which might have arisen had he treated it in the same manner as he did inheritance. The greatest strength of his analysis was that it served to interrelate a considerable body and wide diversity of phenomena, giving it a sound factual basis. Its significance may be particularly well appreciated when we consider the example of how Darwin's thinking on recapitulation fitted in with that of variation. Although Darwin by no means originated the concept of recapitulation, it was he who first gave it its proper theoretical interpretation: "The explanation is, that variations do not necessarily or generally occur at a very early period of embryonic growth, and that such variations are inherited at a corresponding age. As a consequence of this the embryo, even after the parent-form has undergone great modification, is left only slightly modified; and the embryos of widely-different animals which are descended from a common progenitor remain in many important respects like one another and probably like their common progenitor." [22] Darwin accomplished far more than a mere explanation of the phenomenon. He used it to explore the mechanisms of evolution and development. When it was seen that the effect of many embryological processes depends upon the stage at which the processes occur, a broad range of other phenomena became intelligible. To Darwin, ontogeny was far more than the cause or the result of phylogeny; both were to be understood as joint implications of the same fundamental principles. By inquiring into the nature of the underlying processes, he could integrate both ontogenetic and phylogenetic thinking into one comprehensive system.

Once he had conceived of development as a process intelligible in strictly mechanistic terms, Darwin could analyze its manner of operation and see how its laws manifest themselves in the course and pattern of evolutionary change. Among the simplest of the insights that emerged from such thinking, and one of the most fruitful, was the idea that differences between organisms often result from differences in the relative growth of their parts. For example, he showed that the asymmetry in the ears of "half-lop" rabbits correlates with many further deformations in the morphology of the skull, including the jaw.[23] As was mentioned in the chapter on barnacle taxonomy, his use of quantitative evidence and his grasp of taxonomic implications foreshadow quite a number of subsequent developments in mathematical biology. He shows remarkable perceptiveness in enumerating his examples of correlated growth; for example: "It can hardly be an accidental coincidence that the two species of Columba, which are destitute of an oil-gland, have an unusual number of tail-feathers, namely 16, and in this respect resemble Fantails." [24] The works of Darwin abound in such comparisons, in some instances elucidating the laws of variation, or, as in barnacle taxonomy, applied to other theoretical questions.

The concept of relative growth is closely allied to that of "homologous variation." [25] The concept of homology, or formal correspondence of parts, has already been discussed in relation to its bearing on systematics. To Darwin, however, homology was more than a morphological or taxonomic concept; he developed its most general implications and gave it a broader, mechanistic interpretation. It is not only descent from a common ancestor that causes organisms to be constructed according to the same "plan," or with equivalent relations between the parts. We have also such relationships as serial homology and the correspondence between the two sides of the body. Drawing largely upon a work by Isidore Geoffroy Saint-Hilaire, *Histoire des Anomalies*, Darwin had much evidence to show that homologous parts all tend to vary as a unit.[26] He was thus able to elaborate upon an idea that goes back to antiquity, that certain parts are basically the same and, therefore, show the same pattern of variation. Darwin argues, for example, that when hands become monstrous, they tend to resemble feet, and conversely. Again, he uses a comparison which may be found in Aristotle, of hooves, horns, and hair, which show correlated variations in pigmentation or pattern.[27] He notes how Virgil advises shepherds to look at the tongues of white sheep to see if a trace of black pigment might reveal a concealed tendency to darkness which could appear in the wool.[28] He extends this generalization to teeth and other structures, which, being of ectodermal origin, are, therefore, homologous; hereditary disorders in all these parts, he points out, tend to coincide.[29]

The utility of Darwin's morphogenetic thinking may likewise be seen in his concept of "analogous variation." This notion has been misleadingly interpreted as "parallel mutation," but as Darwin had no real idea of a genotype, this view must be rejected.[30] Perhaps Darwin was excessively vague when he gave the following definition: "By the expression analogous variation (and it is one that I shall often have occasion to use) I mean a variation occurring in a species or variety which resembles a normal character in another and distinct species or variety." [31] And the following elaboration is a bit misleading: "Analogous variations may arise . . . from two or more forms with a similar constitution having been exposed to similar conditions, —or from one of two forms having reacquired through reversion a character inherited by the other form from their common progenitor,—or from both forms having reverted to the same ancestral character." [32] But the phenotypic nature of the variations, and the embryological, rather than genetic, explanation, are clear when Darwin says: "The fact of the feathers in widely distinct groups having been modified in an analogous manner, no doubt depends primarily on all the feathers having nearly the same structure and manner of development, and consequently tending to vary in the same manner. We often see a tendency to analogous variability in the plumage of our domestic breeds belonging to distinct species. Thus top-knots have appeared in several species." [33] In other words, closely related forms tend to undergo similar changes because, having basic similarities in anatomical structure and in morphogenetic processes, they are able to undergo the same kinds of modifications. Such an embryological channeling of the course

of evolutionary events is one of the valid explanations for parallelism (i.e., the evolution of similar characters in different, yet
closely related lineages).[34] To be sure, the parallel changes depend, if they are to occur at all, upon selection. It is also true, as
Darwin points out, that selection alone could produce many similarities independently in closely related lineages.[35] But there is no
contradiction.

Darwin's search for laws of variation, and his attempt to reduce
such laws to embryology, were manifestly guided by theoretical
insight. For instance, it was a well-known empirical rule that
white cats with blue eyes are usually deaf. Darwin gathered a
number of facts directly and indirectly related to this phenomenon.[36] He learned that a white cat had given birth to a litter of
kittens, some of which were both white and deaf, while others
were normal in both color and hearing. In cats with only one
blue eye, hearing was normal. And in one cat, both hearing and
pigmentation appeared at the age of four months. It was one of
Darwin's most useful theories, albeit one that did not originate
with him, that monstrosities are largely due to arrested development.[37] This notion, a corollary of the idea that the timing of
developmental processes has a profound effect on structure, he
conjoined with his thinking about pleiotropy. He knew that very
young kittens have blue eyes, and was aware of the fact that
variations in the structure of eyes and ears are correlated, having
noted such facts as the frequency of defective vision in deaf-
mutes. He formulated the hypothesis that hearing and pigmentation are in some manner interrelated during normal development.
He experimented by hitting a shovel with a poker near some very
young kittens. As expected, there was no response. This experiment—obviously no mere natural history observation, but the
natural consequence of theoretical reasoning—readily fits in with
the observations on artificial selection which Darwin invoked to
support his ideas on pleiotropy. Should someone try to establish a
breed of white, blue-eyed cats, this would probably involve their
being deaf as well. Darwin laid great stress upon the fact that in
artificial selection, undesired, correlated characteristics often develop in spite of the breeder's efforts.[38]

VARIATION AND THE ENVIRONMENT

We are now in a position to consider how Darwin's developmental thinking relates to the problem of blending. Some advance in our understanding of such matters was made by Vorzimmer, in refuting the widely held notion that an article by Fleeming Jenkin, published in 1867, brought the problem to Darwin's attention.[39] Vorzimmer argues that the problem arose out of the logical process of Darwin's analysis: in the early stages, the problem was solved by isolation, while in later work (*ca.* 1856), it appeared clear that small populations would decrease the probability that an advantageous variant would appear. It is held that in the first edition of *The Origin of Species*, the problem was solved by more variants and a larger number of eliminations. Vorzimmmer concludes, and I think rightly, that Darwin later came to stress the importance of minor differences between individuals because these are more common; but he upholds the usual tradition that habit and use and disuse were invoked to provide a sufficient number of variant forms.[40] The important distinction would seem to be that there are some changes ("single variations") which are large and rare and do not blend, and others ("individual differences") which are small and common and subject to blending. The relative importance of individual differences is in fact held to be much greater in later writings, as may be seen in the following quotation: "When individuals of the same variety, or even of a distinct variety, are allowed freely to intercross, uniformity of character is ultimately acquired. Some few characters, however, are incapable of fusion, but these are unimportant, as they are often of a semi-monstrous nature, and have suddenly appeared." [41] This is no *ad hoc* hypothesis. The facts did seem to indicate that changes in domesticated organisms had been produced by the selection of small variations, and individual differences were obviously very common. We can hardly blame Darwin for not knowing that much of the observed variation was not inherited; and in spite of this complication, his premises were true and his conclusions correct. It is clear that he did not really modify his basic theory in the face of blending, but only decided that blending affects the manner in which selection may act.

The environment came to play additional roles in Darwin's later thinking on evolutionary mechanisms, but its significance in his theories has been grossly misinterpreted. Darwin was not very explicit about the mechanisms by means of which changed conditions bring about variation. He thought that there must be some kind of action on the reproductive system. But it is easy to be misled on this point, for by "reproductive system" he meant the totality of mechanisms producing new individuals, not the more restricted usage to which we are accustomed. The environmental conditions were thought to affect the process of reproduction at some step, but it was not possible to say how. We may observe Darwin's uncertainty in the following quotation: "We can see in a vague manner that, when the organised and nutrient fluids of the body are not used during growth, or by the wear and tear of the tissues, they will be in excess; and as growth, nutrition, and reproduction are intimately allied processes, this superfluity might disturb the due and proper action of the reproductive organs, and consequently affect the character of the future offspring." [42]

SEX, VARIATION, AND EMBRYOLOGY

One might think that Darwin's interest in the relationship between the reproductive system and variation arose from a desire to formulate a mechanism whereby the conditions of life could affect the hereditary material. It is incontestable that he did elaborate such a theory: the "provisional hypothesis of pangenesis." But Darwin's hypothesizing can better be understood as motivated by attempts to explain, not inheritance, but variation. The environment was to act upon development rather than on heredity. Some of these relations are clear from his *The Effects of Self and Cross Fertilisation in the Vegetable Kingdom,* which treats a long series of experimental studies. The experiments were originally designed to test the little-verified opinions of animal breeders that long-continued inbreeding results in degeneration. As the effects of inbreeding in animals are not immediately obvious, Darwin reasoned that self-fertilization in plants would provide a rigorous test because in plants a greater degree of inbreeding is possible than in domestic animals, which at best can be crossed with close relatives. He asserts that he himself was surprised with the results, for there was a striking difference, on the whole, between self-fertilized and cross-fertilized plants. The progeny resulting from self-fertilization were distinctly inferior in fertility and size.[43] The fact that Darwin did not expect such a difference is most significant, for he was forced to rethink his basic premises and to design additional experiments.

Darwin's selfing experiments were quite straightforward and well designed. He grew, under identical conditions, seeds from plants which had been self-fertilized or cross-fertilized and measured the growth and fertility of the progeny. It may be observed in passing that Darwin's work was subjected to a statistical analysis by his cousin, Francis Galton, one of the founders of biological statistics, and in this sense constitutes something of a novelty.[44] Upon finding that cross-fertilized plants usually give rise to more vigorous and fertile progeny than do self-fertilized ones, Darwin set out to analyze the effects of various possible causes for the difference. A more or less obvious question was whether crossing, as such, had been responsible for the greater vigor of the crossed plants. He demonstrated that it had not, by showing that flowers from the same plant (which have the same hereditary makeup) produce no more vigorous or fertile progeny when intercrossed than when a single flower is allowed to fertilize itself.[45] He also showed that the effect on fertility is not strictly correlated with vigor, suggesting that two different mechanisms are involved.[46] He therefore concluded that the ill effect of inbreeding depends upon an advantage that results from having diverse parents. His inference bears a striking resemblance, at least superficially, to such modern theories as heterosis which are now invoked to explain the advantages of outbreeding.[47] Such parallels are always difficult to interpret, but it should be noted that his explanation interrelated the same sort of phenomena as are now grouped according to the analogous principle. And Darwin was able to integrate this particular generalization into the structure of his other theoretical systems. In view of how little was known of cytology at the time it was formulated, the following inference is most impressive: ". . . that the advantage of a cross depends wholly on the plants differing somewhat in constitution; and that the disadvantages of self-fertilization depend on the two parents, which are combined in the same hermaphrodite flower, having a closely similar constitution. A certain amount of differentiation in the sexual elements seems indispensable for the full fertility of the parents, and for the full vigour of the offspring." [48] Darwin's notion of "sexual elements" is of course quite vague, and his usage is apt to be misleading. But he does seem to have formulated the idea that intrinsic developmental mechanisms are responsible for the vigor and fertility of the offspring. Further, the properties of the progeny are conceived of as something more than just a blend of parental attributes. And he has supported the thesis that developmental mechanisms are a key to an understanding of variation as a process. Hence it seems evident that when he speaks of organisms coming under the same conditions and attaining uniformity of character, he does not concern himself with inheritance, but only asserts that the phenotype becomes uniform. He says nothing about the variability, or other properties, of the underlying mechanisms.

Darwin looked upon outbreeding as a cause of variation, and in his way of thinking developmental processes were of great importance in giving rise to it. He compares outbreeding to the stirring of a fire—a simile connoting an action, not upon characters, but upon that which produces them.[49] His idea of "latent" characters relates to the same kind of developmental thinking. He recognized that the capacity to produce a trait may long remain unrealized, or latent. This notion of latency bears some resemblance to the modern concept of recessive genes—but without the genes. Some of the explanations which he proposes for this phenomenon are rather difficult to reconcile with the idea that all inheritance is of the blending type. He did, in fact, make some distinction between genotype and phenotype. But instead of a material genotype, he provided a vague potentiality or tendency to vary in a particular direction. This is clear from his terminology: *variation* is the process of phenotypic change; *inheritance* is the phenotypic development of a character because it was present in an ancestor; *transmission* is the derivation of a potentiality ("genotype") from some other individual. A trait could remain latent, not because, as modern genetics explains it, its gene was there all the time though unexpressed, but rather because the structure of the developmental mechanism could easily be changed so as to produce it. In a sense, the phenotype could be said to have a certain lability, such that it could be modified into a variety of states, and latency was the ability to realize, under the proper conditions, any of these alternatives. Closely related organisms would vary in the same direction because they are most easily modified in only a limited number of ways (and this is the sort of reasoning that underlies some modern explanations for parallelism). To those unaccumstomed to thinking in such terms, this way of dealing with the problem may seem bizarre, but it explains much. Consider, for example, how Darwin observes that crossing of two canaries with top-knots produces offspring which are bald, or which have wounds on their heads. A simple blending theory would be inconsistent with such an occurrence. But in terms of transmitted effects on developmental mechanisms, one might expect that the cumulative action of those processes which give rise to the property could, under some circumstances, have additive effects. In remarking upon this phenomenon Darwin says: "It would appear as if the top-knot were due to some morbid condition, which is increased to an injurious degree when two birds in this state are paired." [50]

Darwin's strongest argument against the more simpleminded idea of blending inheritance is to be found in his treatment of sexual dimorphism.[51] A cross between a male and a female of the same species gives rise to males and females, not to hermaphrodites. He maintains that the development of sexual characteristics depends, not upon the transmission or nontransmission of a particular set of characteristics, but rather upon embryological action causing the expression of one or another set of characteristics, both of which exist, potentially, in all members of both sexes. That sexual characteristics may remain "latent" he verified by enumerating the effects of castration and diseased gonads, and by noting how old hens come to resemble roosters and how male chicken-pheasant hybrids or castrated cocks may incubate eggs.

The same kind of analysis may be found in Darwin's treatment of atavism (i.e., reversion to an ancestral state). He provides many examples of this phenomenon, such as the presence of zebra-like stripes in crosses between the horse and the ass.[52] He shows how the atavistic traits are often absent in the young, but gradually become more pronounced as the animal or plant gets older— an observation having obvious analogy to the ontogeny of sexual characteristics.[53] Thus it is clear that he conceived of atavism as resulting when modified elements of a developmental mechanism are recombined in such a manner that the original developmental process is reconstituted. The occurrence of atavism in pure lines he thus explains by latency: "If, however, we suppose that every character is derived exclusively from the father or mother, but that many characters lie latent or dormant during a long succession of generations, the foregoing facts are intelligible." [54] This interpretation has a parallel in one modern explanation for some types of atavism. [55] It is widely believed that the ancestral property was produced by a combination of genes, and that the derived conditions have arisen through different modifications of the original pattern. When the derived forms are once again allowed to cross, recombination may produce the ancestral genotype and give rise to the original phenotype as a result. The comparison is imperfect because Darwin did not invoke a genotype, or system of inherited genetic material. But it does make his thinking intelligible, as when he sums up his views on the mechanism of atavism as follows:

> We have seen . . . that when two races or species are crossed there is the strongest tendency to the reappearance in the offspring of long-lost characters, possessed by neither parent nor immediate progenitor. When two white, or red, or black pigeons, of well-established breeds, are united, the offspring are almost sure to inherit the same colours; but when differently-coloured birds are crossed, the opposed forces of inheritance apparently counteract each other, and the tendency which is inherent in both parents to produce slaty-blue offspring becomes predominant. So it is in several other cases. But when, for instance, the ass is crossed with *E. indicus* or with the horse,—animals which have not striped legs, —and the hybrids have conspicuous stripes on their legs and even on their faces, all that can be said is, that an inherent tendency to reversion is evolved through some disturbance in the organization caused by the act of crossing. [56]

It again seems clear that a characteristic is not lost forever through blending, but can be evoked, under certain conditions, from its latent state. [57]

Much the same kind of thinking is to be seen in discussions of what Darwin calls "prepotency." A parallel might be drawn between prepotency and dominance, but this would be misleading, as may be seen from the following statement: "In some cases prepotency apparently depends on the same character being present and visible in one of the two breeds which are crossed, and latent or invisible in the other breed; and in this case it is natural that the character which is potentially present in both breeds should be prepotent." [58] Clearly what is meant is not: $Aa \times Aa$; F_1 $1/4AA$, $1/2Aa$, $1/4aa$, for this implies the "unmasking" of a weak ability to produce a character when the dominant gene is no longer present. Rather, it means that the potentiality to produce a character has an additive effect. Perhaps the closest modern term for "prepotency" is "penetrance." The best analogy is with dominant semilethals, which produce a weak effect in the heterozygous condition and a much stronger one in the homozygous; an example would be the canaries mentioned above, which have top-knots and give rise to bald offspring. But once again, one should remember that no genotype is presupposed.

MODIFICATIONS OF THE THEORY TO ACCOMMODATE BLENDING

Reasoning from his ideas on the process of variation, Darwin was able to suggest conditions under which selection could act in spite of blending, and without having to invoke inherited habit, use and disuse, or other rationalizations. Of primary importance was a model involving environmentally evoked variation combined with a "tendency to vary" in one direction. Suppose that the conditions of life affect developmental processes, that such changes are inherited, and that the changes which take place under the new conditions are similar in closely related organisms. If all these premises were true, a population of organisms brought into new surroundings would tend to undergo the same kind of variation in all its members. The characters so produced would, therefore, be very common, and blending would not swamp them out; selection would then be effective. Those familiar with modern genetics will object that this implies an impossibly high mutation rate. But Darwin did not think in terms of modern genetics, and was referring only to the phenotype. To persist in the use of contemporary idiom, the model did not propose that with the shift in habitat "new genes" are produced, but rather that already existing "genes" become expressed to a different degree. Now, granting that this model does not correspond very closely to reality, we may observe that were its premises true, it would provide a fully adequate answer to the problems supposedly raised by blending. It is even more effective when we see that, through pleiotropy, a host of new modifications in the morphogenetic systems would arise with each shift in conditions, providing a continuous source of new potentialities for variation. One may see the kind of process Darwin had in mind from the following passage:

> We know, also, that the horses taken to the Falkland Islands have, during successive generations, become smaller and weaker, whilst those which have run wild on the Pampas have acquired larger and coarser heads; and such changes are manifestly due, not to any one pair, but to all the individuals having been subjected to the same conditions, aided, perhaps by the principle of reversion. The new sub-breeds in such cases are not descended from any single pair, but from many individuals which have varied in different degrees, but in the same general manner; and we may conclude that the races of man have been similarly produced, the modifications being either the direct result of exposure to different conditions, or the indirect result of some form of selection.[59]

Again, one must remember that, to Darwin, change in conditions giving rise to variation was a brute fact supported by a wealth of empirical evidence; it was not something added for the purpose of saving his hypothesis. The allegation that he adopted environmental causes of variation in the face of criticism, or that he significantly altered his theory late in life, can easily be refuted by the evidence of his preliminary "Sketch" of 1842 and "Essay" of 1844, however true it may be that there was a small change in emphasis.[60] Likewise, the environmental action, as he saw it, does not evoke an adaptive change, but is as fortuitous as the mutations of modern biology. As a final qualification, we may observe that Darwin never came to precise conclusions as to the relative importance of different causes of evolutionary change—beyond his unwavering opinion that selection is of primary importance.

USE AND DISUSE: PANGENESIS

There is no empirical evidence for the notion that Darwin adopted the inheritance of acquired characters because it overcomes the problem of blending. While it is true that use and disuse, inherited habit, and other "Lamarckian" notions would have this effect, other subsidiary hypotheses could, and did, do the same. Without exception, all of Darwin's nonselective evolutionary mechanisms may be shown to have been invoked for other reasons. To understand what his motives were, it is necessary to reconstruct his precise views on the nature of heredity, and this is no easy task. Not only are his ideas somewhat vague and seemingly contradictory, but his manner of expressing himself changes with the kind of problem under discussion. Often his statements are not expressive of his real views, but are merely conjectural. Thus, in *The Descent of Man*, where, in suggesting that mental exercise might benefit the female intellect, he seems to be arguing for the inheritance of acquired characteristics, a closer reading shows that he merely states that such a change would result if, and only if, there is such a type of inheritance.[61] In countless instances the same kind of speculation crops up; he suggests a nonselective mechanism, and even admits it, but he does not conclude how important selection has been. The situation is readily understood, for Darwin theorized at great length about evolution in general, in addition to the mechanism; when any cause would do, he did not deal with the problem.

Faced with the problem of distinguishing between Darwin's speculations and his definitive views, one may have recourse to the conceptual analysis of his theoretical constructs. Particularly useful insights may be gained by considering the effects of his embryological models on his arguments. His really precise and definite conclusions are, on the whole, those which he could derive as implications from his morphogenetic theories. It is revealing to note that if one groups his generalizations about inheritance according to whether they do, or do not, follow from his embryology, it is those basically in accord with his models that have stood up to subsequent investigation. Thus his ideas about correlated growth and variation, and about latency and throwbacks, which derive from developmental hypotheses supported by an enormous quantity of empirical data, retain much validity even today. By contrast, the idea of blending resulted from a superficial comparison; it was never supported by truly critical tests, and it was founded on no theory. Likewise the effects of use and disuse are not implicit in any theory, were never subjected to rigorous testing, and often occur in Darwin's writings as merely plausible reasons for phenomena equally well explained by selection.

Although Darwin accepted the idea that use and disuse affect the course of evolution, he did so on purely empirical grounds: he reasoned that a number of brute facts cannot be otherwise explained. To be sure, his facts were few, and they vanished in the light of further investigation. The crucial experiments which he cites are a series by Brown Séquard, who, for example, claimed to have observed that the amputation of toes in guinea pigs brings about comparable deformities in the young.[62] The bulk of Darwin's further evidence is anecdotal and not of the sort that modern biologists consider reliable. For instance, he accepted the folk-belief that female animals are affected by mating with one partner, in such a manner that subsequent pairings with other individuals sometimes produce offspring resembling the first partner. He relates a number of supposed instances of mutilated people having engendered children with a similar deformity. His reasons for accepting such evidence are stated quite plainly when he says: "In all cases in which a parent has had an organ injured on one side, and two or more offspring are born with the same organ affected on the same side, the chances against mere coincidence are almost infinitely great." [63] Of course, Darwin was wrong here, and the reason is a defect in his methodology. He had little understanding of statistics, and he had not adequately appreciated the dangers inherent in anecdotal evidence. Much of his work on psychology may be justly criticized on the same grounds. Yet we should not forget that in his day scientists in general were far less sophisticated in such matters than they are now. And he was not without cognizance of the problem in its simpler form, for he went out of his way to reject, as unsound, a large body of similar evidence—such as tailless dogs and cats—as unreliable because possibly due to chance.[64] There is no justification whatsoever for believing that Darwin's mistakes in the choice of evidence were anything but honest.

It is clearly demonstrated in Darwin's works that many evolutionary phenomena are utterly inexplicable in terms of use and disuse, and that their effects are quite limited. He states most explicitly that use and disuse affect *only the size* of an organ, and not its structure.[65] Among his better arguments is one based on the fact that in domestic forms useless structures do not tend to become reduced.[66] Indeed, certain parts, such as the fifth toe in dogs, tend to increase in size. The principle of use and disuse is thus powerless to explain certain well-established facts. The theory of natural selection, on the other hand, would imply that rudiments should be retained, owing to the decreased selection pressure that obtains under artificial conditions, where small savings in energy are offset by superfluity of food. Likewise Darwin's ideas about pleiotropy make the increase in size of rudiments quite intelligible, while the other hypotheses had no such merit.

Darwin's definitive stand on the importance of natural selection is given as follows in the later editions of *The Origin of Species:* "I am convinced that Natural Selection has been the most important, but not the exclusive, means of modification." [67] Darwin modified his views only to admit a few exceptions; but how many, he never chose to say. It is, therefore, quite erroneous to assert, as does one commentator: "Finally, in 1868, Darwin's famous chapter on Pangenesis appeared in his *Variation of Animals and Plants Under Domestication* and in this work Darwin showed that he had developed into a complete Lamarckian." [68] Nor is it correct to say that Lamarck's views were adopted to support the theory of evolution. Although it is true that an explanation of many facts in terms of Lamarckian ideas would be consistent with the fact of evolution, and that modification of Darwin's theory might win some converts to evolution, it does not follow that Darwin so modified his theory, or that he altered it for that reason. On the contrary, Darwin's purportedly Lamarckian "provisional hypothesis of pangenesis" was never more than a speculation designed to explain a number of empirical observations.[69] When it is understood what these observations were, it is obvious why, and how, Darwin changed his opinions.

Pangenesis involves the idea that each part of the body gives off minute particles, called "gemmules," which ultimately find

their way to the reproductive system.[70] Here they are assembled
and passed on to the offspring, there to control the course of
development. This reasonably straightforward hypothesis explains many facts in addition to what seemed to be facts at the
time. In the first place, it provided a simple mechanism for both
heredity and generation. It also allowed for the mixture of properties in sexual reproduction, yet was fully consistent with the
facts of asexual reproduction. By the latency of gemmules, it
explained how there could be sexual dimorphism and reversions.
The flow of gemmules through the vegetative tissues could provide mechanisms for such phenomena as bud hybridization and
the action of pollen on the ovary. A stock of gemmules throughout the body would account for regeneration. Finally, a movement of modified gemmules from the somatic tissues to the reproductive organs, where they could be incorporated into the
germ, would allow for the occasional supposed instances of use
and disuse affecting the properties of subsequent generations.

Far from being invoked to support Lamarckianism, pangenesis
was in part designed to refute it. The final disappearance of a
structure which no longer served any function posed a difficulty
for the theory of natural selection. The early stages of reduction
would clearly be advantageous in saving energy and materials. But
such an economy would, it seemed, ultimately become so slight
that the reduction would stop. Early in the development of his
evolutionary thinking, Darwin realized that use and disuse would
explain the ultimate disappearance of the parts and referred to the
views of Lamarck.[71] However, upon further study he saw that
there is no consistent tendency for unused structures to disappear. He attempted, with some success, to bring such phenomena
into line with his embryological thinking.[72] He pointed out that
a part which had tended to vary in one direction would be expected to continue varying in the same manner. He reasoned that
if organisms had such properties that they tended to develop a
structure in an ever decreasing degree, that structure would ultimately disappear, provided that selection did not maintain the
part. Considered apart from Darwin's ideas on embryology, this
seems an *ad hoc* argument, invoking "hypothetical agencies
which control the production of mutations." [73] But such is not
the case; it is fully consistent with both Darwin's thinking and
modern biology to attribute the loss of rudimentary structure to

pleiotropy. A part may disappear because the morphogenetic process which causes its formation is altered by selection for other effects of the same process. In the theory of pangenesis, there is yet another explanation.[74] The gemmules that control the development of a structure could, under certain conditions, themselves undergo degeneration; or else by some accident they might not be transmitted. This explanation, if we substitute "gene" for "gemmule," is remarkably similar to the modern interpretation. The parallels between gemmules and the hereditary material are actually quite extensive.[75]

On the basis of the foregoing arguments, it seems reasonable to conclude that the hypothesis of pangenesis constituted no substantial retreat from natural selection. A need to provide a mechanism for a broad variety of phenomena adequately accounts for the fact that Darwin advanced it. Not only was it a strictly mechanistic theory, but it had other advantages as well, such as furthering the idea of particulate inheritance. Curiously, these positive advantages are usually overlooked when Darwin is castigated merely because his theory was not wholly correct. The features of his hypothesis which are more or less in agreement with modern biology are precisely those—particulate inheritance and latency—which grew out of his embryological thinking. On the other hand, the erroneous features are largely due to bad empirical data. The charge of Lamarckianism, in the sense that the needs or strivings of the organism might evoke appropriate changes in structure, would be, to say the very least, forced.

But why has Darwin been accused of abandoning natural selection, and of becoming a Lamarckian? The answer lies, not with the historical facts, but with the nature of the criticism. In earlier chapters it was shown that many of the accusations have resulted because Darwin did not hold the same erroneous opinions about evolutionary theory as have his critics. It is clear, too, that metaphysical ideas and religious prejudice have detracted from objectivity. To an extent, such error must be indulged, although some would even begrudge Darwin a lesser amount of charity. However, when one finds that there has been no effort at a critical evaluation of the evidence, that facts—when occasionally presented—are frequently given in an inappropriate context, and that outright falsehoods are stated in support of basic theses, while data prejudicial to the argument are conveniently ignored, then not one word can be looked upon as reliable. Can anyone who is aware that Darwin wrote his book on orchids believe anything that is said in works which call Darwin a teleologist? [76] It is hardly conceivable that anyone who had bothered to undertake a careful reading of Darwin's works would assert the following: "In other words we must distinguish, as Darwin usually failed to do, between the historical question of what has happened and the experimental question of how or why it has happened." [77] Again, reference to the original work provides a categorical refutation of this curious assertion: "In *Animals and Plants* where natural selection had been mentioned the direct effects of climate, habit and use had been added. No evidence of course was quoted, for of evidence there was none." [78]

When an intellectual historian ceases to care whether ideas are true or false, he all but inevitably lapses into triviality. But in history there are well-established canons of evidence which allow us to discover the truth, however ugly it may be. Whether we like it or not, the facts compel us to admit that Darwin made a number of blunders and that these had serious consequences. Yet the same reasoning dictates that we acknowledge success along with failure. Darwin's works retain their intrinsic merit, in spite of the fact that sometimes a problem went unsolved or a datum proved misleading. It seems hardly rational, in treating Darwin's genetics along with the rest of his contributions, to condemn the Copernicus and Newton of biology for not being its Einstein as well.

8. *An Evolutionary Psychology*

It is by no means common knowledge that Darwin devoted a considerable portion of his energy to the study of behavior. Histories of psychology mention *The Expression of the Emotions in Man and Animals,* recognize its historical importance, and often praise it for originality and wealth of observational detail.[1] Yet the relationships of this book to Darwin's other works have not been understood, and the result has been a failure to apprehend many of his basic ideas. Any exposition of Darwin's psychological theories is inadequate unless it considers his works on plant physiology, which are concerned with the underlying mechanisms of very simple behavioral phenomena. Likewise his treatise on earthworms deals at great length with behavioral mechanisms in somewhat more complicated systems. One may trace out the systematic development of a comprehensive system of neurophysiological theory, from its roots in the hypotheses of *The Origin of Species,* to its application as a fundamental component in the arguments of *The Descent of Man.* Without an understanding of this particular continuity, Darwin's later writings are largely unintelligible. Indeed, because the underlying frame of reference has been overlooked, Darwin's work on behavior has been interpreted as little more than miscellaneous observations—the casual study of a Sunday naturalist seeking relaxation from the strains of theoretical controversy.[2]

That *The Expression of the Emotions* has not been well understood is clear from the fact that it was a historical dead end. Nobody took up the train of reasoning and developed it, although the work was widely read, and although it did become an element in various controversies. Of late the work has aroused a certain renewed interest, especially among anthropologists concerned with its implications for nonverbal communication.[3] Likewise, those who study communication in animals have found themselves rediscovering certain of Darwin's basic generalizations.[4] But there is little evidence that the real merits of the work have come to be appreciated.

METHODOLOGICAL PROBLEMS IN PSYCHOLOGY

One reason why Darwin's psychology has not been understood is his tendency to express himself in anthropomorphic terms. A natural inference would be that he really believed in a close correspondence between animal and human mentality. There is some truth in this interpretation, but it has many pitfalls as well. Although Darwin frequently attributed too much of a human element to animal behavior, such error generally diminished as he experimented and criticized his theories. Generalization from one species to another has the uncertainty of all inductive inferences, but this does not exclude, a priori, the use of man as a standard of comparison, either in psychology or in anatomy. But more important with respect to Darwin has been a confusion of his language with his real meaning: there is a world of difference between his metaphorical use of anthropomorphic expressions and the propositions which he actually asserts.[5] Many of his statements are easily read in a sense which obviously was not intended. In the following quotation, for example, one tends to get an impression of extreme anthropomorphism; yet the argument is an explicit discussion of the methodological difficulties inherent in trying to explain animal behavior in human terms:

> Sufficient facts have now been given to shew with what care male birds display their various charms, and this they do with the utmost skill. Whilst preening their feathers, they have frequent opportunities for admiring themselves, and of studying how best to exhibit their beauty. But as all the males of the same species display themselves in exactly the same manner, it appears that actions, at first perhaps intentional, have become instinctive. If so, we ought not to accuse birds of conscious vanity; yet when we see a peacock strutting about, with expanded and quivering tail-feathers, he seems the very emblem of pride and vanity.[6]

It is clearly untenable that Darwin was blissfully unaware of the pitfalls of his methods.[7] His critics, however, have found it easy to cast aspersions by quoting his so-called anthropomorphic statements out of context. The absurdity of such accusations becomes particularly obvious when one refers to some of his statements about plants. In his book on climbing plants, for instance, he refers to branches recoiling with "disgust." [8] Again, he notes that "it was an interesting spectacle to watch the long shoot sweeping this grand circle, night and day, in search of some object round which to twine." [9] It should be added that this statement occurs as an integral part of a strictly physiological explanation; a real belief that plants search consciously is out of the question. Leaving aside the distinct possibility that Darwin said this in an ironical sense, it is obvious that he chose such language for its literary effect. He used everyday terminology to convey precise and definite meanings, with elegance and clarity. For instance, Darwin gives two pictures in which he shows the contrasting appearance of cats under conditions involving precisely opposite kinds of behavior.[10] These are entitled "Cat, savage, and prepared to fight . . . ," and "Cat in an affectionate frame of mind. . . ." Contemporary biologists may regard these captions with amusement, even delight, for the prevailing standards of pedantry are opposed to such expressions, even though nobody ever would take them literally, and even though they could scarcely better express the underlying ideas. One commentator thinks that the second of these pictures, one depicting a cat rubbing against someone's leg, ought to be entitled "cutaneous stimulation." [11] As if cutaneous stimulation were the opposite of a savage attitude!

The attacks on Darwin's language are not altogether separable from criticisms of methodology which, when they are applied in the proper context, are quite valid. Such objections cannot be understood without due consideration for their philosophical basis and for the historical conditions which gave rise to them. A major obstacle to the development of our understanding of behavior has been that psychology has more roots in philosophy than in natural science. The habit of relying upon reason alone may lead to vacuous speculation. There is a most tenuous boundary between observing one's own inner experiences—introspection—and the mere cataloging of attitudes that are emotionally satisfying. There is no logical reason why it should not be possible to apply the scientific method in making controlled and rigorous observations on subjective experience. But attempts to do so are full of difficulty; controls are hard to apply, experiments are not readily repeated, and we find it impossible to compare our own inner experience with that of other species. The problem is intrinsically difficult, and it is not surprising that many investigations of an introspective nature have been criticized for lack of rigor.

When one's methods for studying a particular phenomenon are somewhat deficient in rigor or effectiveness, the difficulty may be overcome by various strategies. One way is to put up with the difficulties and seek to make do with the conventional tools, hoping that quantity of data, accuracy of observation, or careful reasoning will prove adequate. Another is to devise new methods for approaching the problem. Or, one may divert one's attention to a related set of questions for which the means of solution are more readily at hand. All of these techniques have been applied, and all with some success. There are still a number of introspectionists, or those who look upon themselves as such, and there is no reason for rejecting their work. An attempt to elaborate a novel methodology may be seen in Gestalt psychology. The third option has been to study overt behavior exclusively: this is characteristic of the very diverse group of doctrines known as behaviorism.

It must be observed that the various methods just suggested for giving rigor to psychological investigation are by no means mutually exclusive. In many instances the question has been not whether methodology should be improved, but how this should be accomplished. Yet there have been unfortunate excesses. It is one thing to argue that the emotions, in the sense of our subjective experience, are so difficult to study that we can best spend our time on other subjects. Most modern psychologists would probably agree that it makes more sense to study "behavior" than "mind." Such a viewpoint, which may be called *methodological behaviorism*, is an effective way of directing one's efforts toward soluble problems. It may be contrasted with what will here be called *metaphysical behaviorism*, that is, the a priori affirmation of such doctrines as the nonexistence of our subjective experiences, or even of any phenomena associated with overt behavior that cannot be observed directly. It is a denial of our ability to go beyond the immediately obvious and deal with underlying relationships. Thus a metaphysical behaviorist would be one who hypostasizes his methodology so as to make it the grounds of support for an empirical proposition. Such a misuse of ideas—the treatment of heuristic aids as if they were laws of nature—is a very common mistake on the part of scientists. For example, in the reconstruction of phylogenetic histories, it is a valid procedure to seek hints as to ancestral structure in embryonic stages, because these often have not been greatly altered; an insistence, however, that the embryo necessarily resembles an ancestral form is only a profession of faith. Geologists, too, occasionally go too far with uniformitarianism and think that it conflicts with the past existence of ice ages, or of other phenomena which do not occur at present. We can see the same kind of mistake in Watson's *Behaviorism,* the major manifesto of the behaviorist movement. Watson goes so far as to contend that thought is nothing more than talking to one's self, asserting that his reason is theoretical, yet not realizing that it follows only from a reification of his methodological assumptions, and not from any theory in the sense of a set of empirical generalizations treated as hypotheses.[12] But perhaps the worst offense of metaphysical behaviorism has been its a priori rejection of the genetic basis of behavioral events. Watson flatly rejected the idea that inheritance plays any role in determining behavior patterns, evidently because of his cultural predisposition toward human equality.[13] Although it is easy to sympathize with such democratic zeal, its concealment of the truth has not been without its deleterious effects. As Hirsch so aptly, although somewhat excessively, puts it: "Behaviorism's counter-factual dogma blinded the behavioral sciences to the significance of meiosis." [14] Darwin, through study of such phenomena as the "tumbling" of certain strains of pigeons, knew full well that behavior is affected by heredity, and it is hard to see how Watson could have argued otherwise on the basis of the facts.[15] Even though the facts could not be ignored forever, rejecting inheritance for metaphysical reasons served only to hinder the progress of psychology. And it is sobering to observe that democratic societies are every bit as prone to the kind of dogmatism that caused the Soviet Union to reject Mendelian inheritance and to embrace Lysenkoism because the latter, like the Watsonian notion, fitted in better with the prevailing creed.[16]

A metaphysical behaviorism thus arises through an attempt, for perfectly good reasons, to base investigations upon certain kinds of evidence which seem particularly reliable. The mistake comes in assuming that the favored evidence is the only kind applicable, or that the discoveries which result from applying the method are the only valid ones. Though the problem is conspicuous in psychology, it troubles all the sciences. A comparable mistake results from an overzealous use of "operational analysis" in attempts to clarify the meanings of scientific theories. Operational analysis, an attempt to analyze concepts by inquiring into the methods of verification, has been widely misapplied in efforts to define scientific terms. [17] The so-called operational criterion of meaning asserts that propositions have no empirical significance unless one may specify possible conditions for verifying them. This leads to all sorts of misinterpretations and efforts to redefine words so as to make them easily susceptible to operational analysis, or to mistaken assertions that a given proposition is not meaningful because it is not practicable to verify it. At worst, one obtains a warped view of scientific theories as a result of attempts to make concepts fit arbitrary standards of lexicographical convenience. Woodger, for example, held that the word "gene" should be dropped in favor of a formalized system for describing breeding experiments.[18] This maneuver has been widely condemned.[19] In using the word "gene," one implies more than is merely given by the available data; one needs to have some means of communicating his hypotheses. It did in fact turn out that there is no entity strictly corresponding to the gene of old-fashioned "bean-bag" genetics—but progress in science depended upon looking for something like it, and not on merely describing a limited range of experiments. In like manner, attacks on the biological species concept, on the grounds that it is not easy to tell just what populations are species, merely trade a meaningful proposition for a convenient definition of a word.

The main positive value of behaviorism has been that it imposes rigor of methodology by making it easier to refute hypotheses. Yet it was by no means the first attempt at such reform, as may be seen in Lloyd Morgan's canon, first published in 1894: *"In no case may we interpret an action as the outcome of the exercise of a higher psychical faculty, if it can be interpreted as the outcome of the exercise of one which stands lower in the psychological scale."* [20] Nor would anyone who had read Darwin's works with the care which they deserve deny that great pains were taken to maintain rigor of technique; it is the kind of precaution, not the lack of it, that makes his work seem odd. Indeed, the search for a better methodology is characteristic of all good science, as may be seen, for example, in the uniformitarianism of both Lyell and Darwin. But unlike Watson, Lyell and Darwin did not embrace methodological rules to such an extent as to reject obvious empirical facts. One might make a case for the notion that Lyell rejected biological evolution because the stability of species fitted better with his uniformitarianism; but the fact remains that he ultimately did change his mind. If one understands what one is doing, the hypostasizing of methodological rules is not likely to pose a problem. Yet with respect to behaviorism, the difficulty has been considerable. The situation was not impoved by the actions of logical positivists, who entered the battle without sufficient understanding of the biological issues. But with time the original excesses have largely given way to a more sophisticated set of attitudes: probably no psychologist embraces behaviorism in its original form.[21]

DARWIN'S APPROACH TO PSYCHOLOGY: HIS WORK ON PLANTS

Criticisms of Darwin's psychology are largely due, then, to misunderstanding about his methodology, especially to erroneous judgments as to the rigor of his verification. A proper appreciation must rest upon the intrinsic merits of his work, as far as possible apart from the preconceptions of any school. In order to discover and evaluate his real views, we shall need a valid reconstruction of his arguments. Once this has been obtained, it should be clear wherein lie the real merits and shortcomings of the theories in question.

Fundamental to the present analysis is the fact that Darwin's approach to behavioral problems displays many similarities to his work on embryology. Indeed, the two cannot be separated, for his study of plant behavior was based upon movements which result from growth. Darwin's ability to grasp a soluble problem is apparent in his approach to behavior by way of experiments upon very simple systems, such as the responses of specialized leaves to chemical stimuli, and the simple reactions in the growing tips of leaves and roots. He was able to formulate profoundly significant questions concerning the manner in which a combination of sensitivity and controlled growth could impose adaptively beneficial order on the movements of plants. In a manner typical of his biological investigations, he combined observation and experiment with the elaboration of evolutionary hypotheses. By showing a mutual coherence beween hypothesized physiological mechanisms and historical order, he was, as usual, able to verify both kinds of generalization.

Perhaps the simplest of Darwin's behavioral works is his book on climbing plants, a comparative study on the various mechanisms which allow vines to grow upward without having to provide their own support.[22] Darwin showed that the growing tip of a vine moves in a circular path, sweeping out a large area as it goes. As a result of this movement, it is highly probable that the shoot will come into contact with some sort of upright stalk or other object. Once it meets with an object of the proper size and shape, a continued circling movement around the support would result in the vine wrapping itself into a tight coil and thus obtaining a means of attachment. The process may be compared to the way in which a weighted string, when swung in a circle, will wrap itself around any post or stick it happens to contact. The bolas of the Argentinean gaucho operate according to this principle; and in the *Naturalist's Voyage*, Darwin describes at some length the way the gaucho releases a set of weights connected with thongs, and how they wind about the legs of an emu or a horse.[23] Whether these observations provided a heuristic aid is conjectural. But we may note that there is a clear-cut formal similarity in the hypothesized mechanism to the coral reef theory, and even to natural selection: a restriction of random movements provides a directional effect. "Selective-retention" models typify Darwin's style of reasoning. Darwin goes on to show how it is possible, simply by modifying a preexisting circling movement, to produce the various known mechanisms of climbing in plants. He verifies his hypothesis by a phylogenetic survey of the various climbing plants, showing how, for example, tendril-climbers and leaf-climbers have properties readily explicable as derivative of an earlier, simple circling movement; but the specialized types of mechanisms are not readily derived from each other.[24] This comparison results in a coherent, explanatory system, interrelating evolutionary theory, systematics, and plant physiology; it tests propositions about all these fields.

Although such matters are always debatable, Darwin may be considered the founder of comparative psychology—comparative, not in the sense that he studied rats as well as men, but that he incorporated premises about the differences between organisms into the structure of his hypothetico-deductive systems and so obtained a means of testing his theories.[25] He used this same approach—tracing out the variations in behavioral phenomena through the genealogical tree—in his work on insectivorous plants. In *The Origin of Species* he treats the slave-making habits of ants and the manner of comb formation among bees in essentially the same fashion.[26] It was altogether natural for the author of the *Monograph on the Sub-class Cirripedia* to test his physiological and evolutionary psychology by means of phylogenetic comparisons, just as he tested his notions on developmental mechanics and sex.

In his work on climbing plants, Darwin did not explain the origin of the circling movement. As he treated variation in his argument for natural selection, so he treated such movement as a brute fact. However, he later on extended his theory to provide an explanation for all plant movements, including those of climbing plants. This explanation, developed at great length in *The Power of Movement in Plants*, provides the basic starting point. The impact of this book on the field of plant physiology can scarcely be exaggerated; it is the direct historical precursor of a vast field of research on plant growth substances and developmental mechanisms. The reason for its success is obvious: here, in contrast to many other fields in which he worked, Darwin had enthusiastic supporters in his son Francis Darwin (1848–1925), the German botanist Wilhelm Pfeffer (1845–1920), and Pfeffer's students.[27] Yet because the later development of Darwin's findings was largely carried out by experimental physiologists, the broader meaning of the work has received inadequate recognition. Darwin's arguments are quite clear, but only if viewed from the perspective of his comparative methodology; and the fact that his theories were long controversial is perhaps best attributed to the failure of other workers to treat his experiments in the proper context. Even Francis Darwin, who assisted in the experiments discussed in *The Power of Movement in Plants,* did not fully understand his father's system. This is clear from Francis Darwin's historical account published in 1909, in which the invalid criticisms given by physiologists of the time are repeated, while the intricate anatomical and evolutionary thinking is ignored.[28] But one should not be surprised at this oversight. The younger Darwin was a good experimental physiologist, yet he did not habitually reason in terms of comparative anatomy or evolution. It is characteristic of many great physiologists, such as Darwin and Harvey, that they can combine the comparative approach with the experimental.

The idea giving evolutionary unity to *The Power of Movement in Plants* is that the generality of movements are derived from a precursor called "circumnutation." Darwin shows, by a long series of controlled observations, that the growing parts of plants—root tips, young leaves, apices of shoots—undergo a slight but perceptible movement in a circle. Although the circular movement may often reverse itself, it is continuous. Darwin reasoned that such a circular movement could be modified in diverse ways, merely by superimposing a stimulus-response system. Thus, a growth in the direction of light (positive phototropism) could be produced by accentuating movement as the light source was approached, or by retarding it as the part moved away from the light. Once again, we may perceive a formal similarity to the coral reef theory, in that a new influence comes to impose directionality upon random movements. And once more, too, there is no need to explain what actually causes a part to circumnutate. Much confusion would seem to have arisen because certain physiologists attempted to explain circumnutation as the result of responses to gravity, as if there were a contradiction.[29]

When the circumnutation theory is considered apart from its unnecessary components, one may observe that what is modified is not just circumnutation, but also the processes of growth which produce it. As an implication, there is no need to invoke a simultaneous origin of the movement and of the means for controlling it. This premise, basic to the entire argument, would seem to have attracted little attention.[30] In the theory of natural selection it is implicit that behavioral mechanisms, like all biological entities, must be elaborated by gradual steps, each of which is somehow functional. Should sensation come first, without movement, the former would have no selective advantage. If, on the contrary, an adaptive or purely accidental movement without sensory control were the original stage, natural selection could readily improve upon it by gradually superimposing a stimulus-response system. And such a model could be verified, for it predicts that the physiology of growth in response to stimuli will be modified in particular ways, and that the resulting diversity should therefore correlate with the structure of the classification system. By analyzing the possible causes of evolutionary trends, Darwin could discover the underlying physiological mechanisms. And armed with a few sound hypotheses, he could perform further experiments and observations to extend his system. In essence he was using the concepts of evolutionary theory as tools for investigating development and behavior. Viewed in this light, the often repeated notion that Darwin took up his botanical studies for rest and relaxation during his labor in amassing facts to support evolution, seems a trifle ludicrous.[31] As his successes amply demonstrate, Darwin studied botany largely because plants are ideal materials for solving certain kinds of problems.

It is difficult to make out the precise sequence of steps in Darwin's reasoning on the mechanisms of plant behavioral physiology. However, a few examples should illustrate his general approach. We may begin with a consideration of Darwin's objections to the theory of Sachs, that all movement is due to differential growth on the convex side of the bending part—a view which fairly well accords with the modern interpretation. Darwin admitted that the phenomenon of circumnutation is consistent with Sachs's hypothesis of differential growth, but he went on to observe that in response to touch, the reaction is so fast that it can hardly be explained as a growth response.[32] In his works on climbing and insectivorous plants, Darwin argued that in some instances the concave side may contract. In *The Power of Movement in Plants*, he draws attention to the pulvini—joint-like regions in the petioles of many leaves, which allow reversible movements.[33] He observes that these areas consist of small cells, and compares the histology of pulvini to that of embryonic tissues, explaining their origin by developmental arrest, which, as we have seen, he considered a major mechanism of evolutionary change.[34] He argues that the small size of the cells facilitates the actions of expansion and contraction, through the resulting increase in the relative area of the walls.[35] Comparative considerations thus would seem to have played a very important role in Darwin's thinking on the mechanisms of movement.

Darwin's work on tropisms was highly controversial, although it has ultimately received much corroboration. Sachs maintained that the response to light results from the direct action of light on the cells in the area where bending occurs, with the light inhibiting growth on the side toward the light source, so that greater expansion of tissue on the far side would make the plant grow toward the light.[36] It is by no means certain how Darwin arrived at his own explanation, but it seems a reasonable conjecture that he did so as a consequence of his theorizing about the comparative physiology of directed growth in general. If he were seeking an explanation that would cover such responses as thigmotropism, as well as those to light, he needed something more elaborate. Darwin ultimately hypothesized that light acts indirectly, at some distance from the area of differential growth. This was shown to be the case by a series of classical experiments involving such manipulations as painting various areas and seeing if there were any effect attributable to diminished light. In what is perhaps the most celebrated of these experiments, he covered up the growing tip in such structures as the oat coleoptile and showed that doing so abolished the curvature which, in controls, occurred at some distance from the tip.[37] The fact that an equal amount of light reaches the cells potentially able to produce the curvature in both experimentally treated plants and controls demonstrated that the response did not result from the direct action of light at the place of bending.

Darwin probably obtained his insights on the reactions of plants to distant stimulation as a consequence of observations on orchids and insectivorous plants.[38] He remarks in his autobiography that in 1860 he first noticed insects on the leaves of *Drosera* and, upon seeing the leaves move, suspected that these motions had some adaptive significance. As he puts it: "Fortunately a critical test occurred to me, that of placing a large number of leaves in various nitrogenous and non-nitrogenous fluids of equal density; and as I found that the former alone excited energetic movements, it was obvious that here was a fine new field for investigation." [39] Darwin's attempts at comparing the physiology of plants to that of animals can be traced to this apparent community of nutritive function. After a long series of experiments on the plants' digestive fluid, he concluded that its action is comparable to that of the mammalian enzyme pepsin.[40] He carried his investigation so far as to test the responses of various plants to substances known to act upon mammalian nervous systems, including alkaloids and narcotics.[41] He demonstrated that the kind of response depends upon the chemical nature of the stimulating substance, but found no close analogy to effects on animal tissues.[42] By observing the sequence of bending, and by following the visible manifestations of some cytological events which accompany the transmission of stimuli, he followed the course of events which lead to reaction at some distance from the point of stimulation.[43] We are fortunate that Darwin himself related the history of how this insight was extended to elucidate the mechanims of transmission, for it gives a most straightforward example of his manner of working.[44] He found that the reaction of the hairs on the surface of the sundew leaf occurs in a sort of wave, with one series of hairs after another beginning to bend; but the wave of transmission occurs more rapidly along the long axis of the leaf than in the transverse direction.[45] A look at the orientation of vascular bundles showed that these run in the direction of most rapid transmission. It occurred to him that the impulse might pass along the vascular tissue. He severed the bundles in various places and found some confirmation. We can see his enthusiasm, as well as his goal of a comprehensive theory of behavior for both plants and animals, when, in a letter of 1872 to Asa Gray, he writes: "The point which has interested me most is tracing the *nerves!* which follow the vascular bundles. By a prick with a sharp lancet at a certain point, I can paralyse one half the leaf, so that a stimulus to the other half causes no movement. It is just like dividing the spinal marrow of a frog: —no stimulus can be sent from the brain or anterior part of the spine to the hind legs; but if these latter are stimulated, they move by reflex action." [46] Later, a more detailed study led to a few contradictions. Therefore, Darwin rejected his hypothesis and attributed the more rapid longitudinal transmission to the arrangement of the cells. The cells are laid out with their long axes parallel to the axis of the leaf, and the impulse is slowed down when passing transversely because it must cross a larger number of cells when traveling in that direction.

When one views Darwin's work on responses to stimuli as a coherent unit, it may be seen that he designed his experiments with a clearly formulated body of questions about the mechanisms of plant "neurophysiology" in general. His explanations for the various tropisms all derive from the hypothesis of a fundamental community of mechanism responsible for all plant movements. He extended his experimentation to cover a variety of responses, finding support for his ideas (some of which have since been refuted) that a reception of the stimulus at the tips of roots and a subsequent growth more distally bring about such responses as hygrotropism, geotropism, and thigmotropism.[47] All these movements, he concluded, could be referred to the same kind of circumnutation that underlies the directed growth of shoots and leaves. Darwin's theory of circumnutation was, then, no mere speculation: it was an indispensable element in a comprehensive system of plant sensory and motor physiology. Stress was laid upon efforts to demonstrate circumnutation, not to lend subjective plausibility to the theory, but to unify the separate parts of the system, by giving concrete evidence that both roots and tendrils are controlled by the same basic mechanism. From the fact that the various organs of plants behave in a manner predicted by a single set of hypotheses, from his anatomical insights, and from the various kinds of evolutionary changes known to have occurred, he made a number of sweeping generalizations. The broad outlines of his theoretical system are tenable in the light of contemporary knowledge. One might read in any modern textbook of plant physiology how various responses of plants to stimuli are mediated by the action of plant hormones which operate in much the same fashion wherever they occur. A major reason why Darwin's experiments were criticized is that the basic structure of his theoretical system was not understood: his manner of reasoning was alien to conventional physiology.

From his experimental work, Darwin was able to show many analogies between the behavior of plants and that of animals.[48] Evidently, he was seeking to lay the foundations for a comprehensive, evolutionary theory of neural integration. In the summary chapter of *The Power of Movement in Plants*, he stresses both gradual evolution and the analogy with animals.[49] Indeed, he goes so far as to compare the root-tip, with its ability to react in an adaptive manner in response to a variety of stimuli, to a brain.[50] Nonetheless, he took pains to draw a sharp distinction between the two kingdoms in the nature of the parts which make up the integrative apparatus. In our age of cybernetics, such comparisons have become commonplace.

BEHAVIORAL STUDIES ON SIMPLE ANIMALS

The problem of the origin of higher mental faculties is taken up on a somewhat more ambitious scale in Darwin's celebrated book on worms. This, the last of his books (1881), was the product of extensive study over many years, going back to a paper of 1837.[51] His interest in behavior is a dominant theme in this work, which in a sense fills the gap between his studies on plants and man with a series of observations on organisms of intermediate complexity. To be sure, the work is not purely psychological. It deals with a broad range of ecological phenomena, treating the role of worms in changing the landscape, forming soil, preserving old ruins, and in various geological effects. It even has a philosophical message, which is one reason why it sold exceedingly well; it exalts the humble, industrious worm to the status of a major geological force.

Much light is cast upon Darwin's attitudes toward a number of psychological concepts by his treatment of the "mental qualities" of worms. He looked upon the earthworm as manifesting a very early stage in the development of "intelligence." By this term he did not mean, as one might think, to draw an anthropomorphic comparison with abstract thought. In a letter to G. J. Romanes (1848–1894), one of the founders of comparative psychology, he defends his use of the term: "I have not attempted to define intelligence; but have quoted your remarks on experience, and have shown how far they apply to worms. It seems to me that they must be said to work with some intelligence, anyhow they are not guided by a blind instinct." [52] It is typical of Darwin's means of expressing himself that he does not bother to define his terms any more than is necessary to allow his readers to make the distinction he has in mind. In this case, he is concerned with the difference between strictly innate behavior patterns and those involving some kind of adjustment, such as learning. He held, further, that simple reflex actions must not be confused with behavior involving a considerable amount of integrative activity. His reasons for believing that a worm's behavior is more elaborate than a set of simple reflexes are of some interest. He observes that when worms are active at some task, their sensitivity to minute changes in light intensity decreases; this is compared to attention in higher organisms.[53] If one does not wrench this generalization from its proper context, it holds up reasonably well, as it only asserts that worms behave in a manner suggesting a fair degree of complexity in the structure and function of their central nervous systems. He goes on to discuss more complicated behavior patterns, which serve as the basis for his inference that worms "exhibit some intelligence." [54] To attribute "higher" mental processes to "lower" organisms was one way of arguing for evolution. Yet we may see that he is not exalting the mental capacities of worms, but is seeking to degrade the "higher" animals. Observing that worms habitually drag leaves into their burrows by the tips, he realized that this is the mechanically advantageous place to grasp them: the leaves, being pointed, are most easily stuffed into the narrow burrows tip first. He tried bunches of pine needles, remarking that no pine trees grew in his area, and found that these were drawn in by the base.[55] Again, a bundle of needles is most easily drawn into a hole in one particular direction, and the worms grasped the most advantageous end. But Darwin rejected the hypothesis that worms form a "general notion" about ease of manipulation, on the grounds of experiments in which he so modified the needles that they could be drawn in equally well either way; yet the base was still used.[56] Reasoning from the basic idea that the ease of drawing a leaf into the burrow depends upon shape, Darwin constructed a number of artificial "leaves." These consisted of elongate, paper triangles, which he left near the burrows, under both laboratory and field conditions. He found that the worms consistently preferred to seize the triangles by the tips, although they occasionally grasped the sides, or still less often the bases. He took such precautions as seeing if marks left on the paper by the worms might indicate that the experimental results were prejudiced by greater success when the triangles are grasped in the most effective way.[57] From his experiments and observations, he concluded that

"We may therefore infer—improbable as is the inference—that worms are able by some means to judge which is the best end by which to draw triangles of paper into their burrows."[58] Other workers have repeated and confirmed these experiments, although more extensive study has suggested differing interpretations. By a series of experiments involving trimmed leaves and paper constructions, Hanel showed that under some conditions earthworms will seize the leaves in a disadvantageous manner.[59] Others have attempted to demonstrate that trial and error, or perhaps ease in grasping the leaf in particular areas, will account for the observed phenomena.[60] It has been known for many years that earthworms are capable of learning.[61] But the whole problem of interpreting the conflicting results and hypotheses of various authors becomes most difficult when it is realized that the behavior of different species is not necessarily the same.[62] The problem has thus not been solved, but more recent work does support Darwin's view that the ostensibly simple nervous system of the earthworm is sufficiently complicated to resist a simpleminded explanation for its activities.

EVOLUTIONARY PSYCHOLOGY OF HIGHER ANIMALS

Behavior is treated at an even higher level in *The Expression of the Emotions in Man and Animals,* one of the least understood of Darwin's works. It is not simply the anthropomorphic terminology, nor the lapses into anecdotal evidence and genuine anthropomorphism, that prevent the reader from understanding Darwin's message. Those who are interested only in certain aspects of the work quite naturally make such mistakes as thinking that it deals more with communication than with the emotions.[63] Although the expression of emotions does frequently result in their communication, this fact has little bearing upon the basic message of the book. And although many have criticized the work on the grounds that it is based upon Lamarckian premises, it still abounds in ideas that stand on their own merits.[64] The reader tends to miss the point mostly because the system must be grasped in terms of its basic premises before the details are comprehensible.

Darwin organizes *The Expression of the Emotions* in terms of three general principles.[65] The first of these is the *principle of serviceable associated habits*. Under this heading he includes those complex groups of actions which are associated with a particular "state of the mind," and which are adaptive in relation to biologically important situations. A cat's tightening of its muscles, unsheathing of its claws, and other movements which prepare it for combat would fit in this category. A second principle is that of *antithesis*. Darwin observed that when, for example, a dog is greeting his master, the animal's actions are opposite in form to those of a dog reacting in a hostile fashion to a stranger. An "angry" dog stiffens his spine, while a "friendly" dog wags his tail—indeed, he may undulate his entire body; the former movements would be serviceable, the latter antithetical. In contrast to the serviceable associated habits, the antithetical patterns of behavior are without adaptive significance as such. A stiff spine is useful in a fight, but the opposite pattern—undulatory motion—has no immediate utility; although it may communicate a "state of the mind," its component movements accomplish nothing. The third principle is that of *direct action of the nervous system*. Here he grouped a variety of behavior patterns which result by accident from the manner in which the nervous system happens to be organized. Among these he classifies the results of insufficiently controlled nervous stimulation (such as trembling) and other biologically inappropriate or useless actions.

The role of "emotion" in Darwin's system has been greatly misunderstood. He made no effort to give this term a precise meaning, and, as he accepted variation in the theory of natural selection, he accepted the existence of emotional states as a posit. He did not mean to imply that our subjective feelings, such as those which we associate with aggression, are necessarily present in other species. The subjective experiences are totally irrelevant to his theory and may be treated as epiphenomena. The basic idea in the system is that certain internal processes are tied together with such visible manifestations as we associate with anger or fear. The interconnecting of the various processes which accompany an "emotional" condition is treated as a strictly neurophysiological fact, with empirical consequences that may be studied by experiment and observation. Darwin was attempting to verify hypotheses about the interrelationships between certain physiological states, traditionally called emotions, and evolutionary and neuromuscular processes. Whether one chooses to call such states "emotions" was a purely semantic issue which he rightly chose to ignore. The kind of approach that Darwin used in attempting to cast some light on the inner workings of the nervous system is quite respectable in many scientific circles and is often referred to as the "black box" technique.[66] One constructs a model of what is perhaps going on inside, and looks for visible manifestations to confirm or refute the hypothesized mechanism. In a sense, the "black box" is a propaganda device: we tend to forget that all support of scientific theories is indirect to some degree.

Certain other aspects of Darwin's work on expression are likewise apt to divert attention from his underlying concepts. It is true that the facial and other expressions are used in communication, and that Darwin took some interest in this sort of effect. Another motive for studying expression was to combat the notion of Bell that certain muscles in man are present solely "for" expressing certain attitudes—this, indeed, seems to have originally attracted his attention to the subject.[67] And Darwin provided a plausible explanation for the historical origin of various behavioral phenomena. But these are not the basic problems with which he was concerned. Rather, *The Expression of the Emotions* deals with the process by which "emotions" come to be expressed, much as *The Variation of Animals and Plants under Domestication* treats the genesis of variations from an embryological point of view. Analogously, too, Darwin was concerned with the effects of that process which brings the expression into being, upon the course of evolutionary change.

The actual details of Darwin's theory of expression bear many striking resemblances to his ideas on the nature of variation. And it is in precisely those aspects in which the two systems correspond that both have been misconstrued. Darwin thought that many behavioral phenomena have resulted through accidents of history comparable to the pleiotropic effects which he discoursed upon at such great length. He did not believe, as many have believed, that all behavior patterns have some adaptive significance, say, as directly serviceable or communicative. The partially nonadaptive nature of many behavioral phenomena is particularly well seen in his principle of direct action.[68] He aimed to show that many visible manifestations of emotion owe their particular configuration to the type of innervation: tears, for instance, are produced when the face is irritated, because the eyes and face share a common nerve supply.[69] Again, he attributes many movements, such as the lashing of a cat's tail, to an incomplete voluntary control over various sets of muscles.[70]

The idea of direct action is fairly readily understood, and that of serviceable associated habits is almost self-explanatory. It is in interpreting anthithesis that the greatest confusion has resulted. Lloyd Morgan, for instance, freely admitted his inability to understand Darwin's position on antithesis.[71] Marler suggests that antithesis is the result of natural selection—that is, that the antithetical patterns are produced because they are advantageous to the organism in conveying attitudes and intentions.[72] There can be no serious doubt that visual communication is very important in the lives of a wide variety of animals, and that natural selection has influenced its evolution. However, in Darwin's system the communicative function of emotional expression was greatly minimized. Darwin had several reasons for de-emphasizing the communicative significance of any particular pattern of expression in bringing about its evolutionary origin. In the first place, many expressions are only incidentally visible: blushing, for instance, may be detected in dark-skinned people.[73] According to Darwin, we blush quite involuntarily; and there are even better examples. For instance, the galvanic skin response can be used to infer someone's emotional state—as in the so-called lie detector; but nobody would be so rash as to assert that this use of it was the selective advantage that caused it to evolve. The mere fact of communication does not suffice to demonstrate the origin of an expression through natural selection; we see that other factors must sometimes be invoked from the fact that in spite of any advantages to the species, natural selection could not have produced sterility. Darwin took the greatest pains to demonstrate that outward signs of emotion are present both when there is some direct utility for the action and when the action is produced by accident.[74] Again, the great intensity of certain antithetical patterns of behavior, such as those with which a dog greets its master, suggest that the degree of activity is not proportional to any communicative effect. And it is often disadvantageous for us to express our inner feelings.

Darwin was able to show that certain types of behavior patterns are without immediate functional significance; instead, they are the result of actions which would be appropriate under similar emotional conditions. Thus, he explains the tendency of a dog to lick its master's face by comparing it to the habit of cleansing puppies with the tongue.[75] Again, cats under conditions where one would expect them to be experiencing pleasure often push with their feet, much as kittens push against their mother's mammary glands when nursing.[76] This parallel between adult and infant behavior suggests an analogy with the concept of regression, which took such a dominant position in the systems of Hughlings Jackson and Freud. The parallel is by no means accidental. In both behavioral and morphological evolution, and in the development of the individual as well, integrated systems are elaborated from some precursor. And modification of the developmental system is particularly likely to bring about a reversion to the original state. The phylogenetic analogue of regression is atavism—the return to ancestral conditions, treated in an earlier chapter. Regressive phenomena in psychology may involve throwbacks to an earlier stage in either ontogeny or phylogeny. For example, in one of his rarely read publications, Darwin notes that when hives of bees lose the queen, the drones are no longer expelled, and suggests that there is a reversion to an earlier state in which several reproductives coexisted in the same hive.[77] One can see that if the impulse for expelling the drones had for some reason evolved so as to depend upon the presence of the queen, her absence would restore the original pattern of behavior. We may here see the true nature of Darwin's analogical reasoning. The parallel between the different models does not exist because the simpler forms of induction happened to be applied to phenomena with similar causes. Rather, a grasp of the underlying causes, and an ability to apply the same model to the same kind of problem, predisposed Darwin to observe the analogous phenomena.

Darwin's explanation for antithesis is essentially a corollary of that for direct action. He held that bodily actions result from patterns of correlated movements in which the whole body acts as a unit. It is this sort of phenomenon to which Darwin refers when he gives the following example: "It appears that when a dog is chased, or in danger of being struck behind, or of anything falling on him, in all these cases he wishes to withdraw as quickly as possible his whole hind-quarters, and that from some sympathy or connection between the muscles, the tail is then drawn closely inwards." [78] In this same class of phenomena, Darwin also invokes what is now called empathy. He compares the action of a spectator in billiards going through unnecessary imitations of the player and also engaging in actions directly opposite to the player's.[79] This he interprets as meaning that for any pattern of activity, with its subservient set of muscular movements, there is an opposite pattern, with antagonistic muscles brought into play. The evocation of the opposing set of neuromuscular activities, mediated by the opposite "state of mind," brings about the antithetical pattern. The parallel with the antagonistic action of the sympathetic and parasympathetic nervous systems is most striking. And once again, it is well to stress the analogy with pleiotropy: certain irrelevant or useless properties arise because of their dependence upon some adaptive process or relationship.

Viewed in the appropriate context, Darwin's psychological system may thus be seen to have much merit. Even if one rejects his explanations because they do not provide laws for the prediction of behavior, one must grant that he did invent a methodology for inferring those properties of the nervous system that must be attributed to historical accident. And theories of behavior which ignore problems of origin can scarcely be called complete. Nonetheless, his work has not received the attention it deserves. Yet his misleading language and his evolutionary orientation seem hardly adequate to explain the oversight, and we know that the work has been widely read. The problem of finding the precise reasons why Darwin's psychological theories have been so ill-understood is by no means simple and is too broad to be treated extensively here. However, a number of plausible explanations deserve at least passing consideration.

One source of confusion has been Darwin's use of the idea of inherited habits, which he often invokes to explain the origin of psychological properties. He thought that a habit might sometimes originate through learning or insight and become automatic, or inherited in later generations.[80] His empirical justification for this notion is stated as follows in *The Descent of Man*: ". . . some intelligent actions, after being performed during several generations, become converted into instincts and are inherited, as when birds on oceanic islands learn to avoid man. These actions may then be said to be degraded in character, for they are no longer performed through reason or from experience." [81] It is clear that Darwin was misled by the rapid development of learned responses, followed by their maintenance through a kind of cultural tradition in social birds. But he goes on to qualify this inference:

> But the greater number of the more complex instincts appear to have been gained in a wholly different manner, through the natural selection of variations of simpler instinctive actions. Such variations appear to arise from the same unknown causes acting on the cerebral organization, which induce slight variations or individual differences in other parts of the body. . . . We can, I think, come to no other conclusion with respect to the origin of the more complex instincts, when we reflect on the marvellous instincts of sterile worker-ants and bees, which leave no offspring to inherit the effects of experience and of modified habits. [82]

Darwin devoted a large section in *The Origin of Species* to the evolution of complex behavior patterns among the neuter casts of social insects. [83] An important motive was to refute inherited habit and Lamarckian mechanisms as sufficient to account for evolution. He writes: "For no amount of exercise, or habit, or volition, in the utterly sterile members of a community could possibly have affected the structure or instincts of the fertile members, which alone leave descendants. I am surprised that no one has advanced this demonstrative case of neuter insects, against the well-known doctrine of Lamarck." [84] But inherited habit does not contradict natural selection, for the two could coexist. Those who criticize *The Expression of the Emotions* on the grounds that it presupposes inherited habit overlook the fact that Darwin invokes natural selection as a more effective mechanism. [85]

Another point at which Darwin's system has been attacked is his notion of instinct. Criticisms of the concept of instinct have come, as one might expect, largely from psychologists, while zoologists on the whole find the term agreeable. [86] The reason for this difference in attitude can be attributed in part to the historical consequences of behaviorism having rejected genetics. The idea of a nature-nurture controversy is simply meaningless to a well-trained biologist. [87] But those who have little understanding of embryology or genetics may actually believe that a trait must be either inherited or not—even though everybody should know that the effects of sunbathing on the phenotype do not contradict the genetic basis of racial differences in skin color. The trouble for many psychologists has been a failure to realize that the learned and unlearned components of behavior may easily co-exist. In his use of "instinct" Darwin merely differentiates between the learned and unlearned components. This he makes clear when he writes:

> I will not attempt any definition of instinct. It would be easy to show that several distinct mental actions are commonly embraced by this term; but every one understands what is meant, when it is said that instinct impells the cuckoo to migrate and to lay her eggs in other birds' nests. An action, which we ourselves should require experience to enable us to perform, when performed by an animal, more especially by a very young one, without any experience, and when performed by many individuals in the same way, without their knowing for what purpose it is performed, is usually said to be instinctive. But I could show that none of these characters of instinct are universal. A little dose, as Pierre Huber expresses it, of judgment or reason, often comes into play, even in animals very low in the scale of nature.[88]

The distinction, then, is in no way absolute. But the fact that one may criticize a word on the grounds that its designatum is not discrete, says nothing about any proposition in which the word occurs. There have been, in psychology, although much less in zoology, numerous misapplications of the term "instinct." The rational solution to the problem is not more talk about the words, but a better grasp of the concepts.

THE FAILURE OF PSYCHOLOGY TO ASSIMILATE DARWINISM

It is easy to see how a psychologist, attempting to give evolutionary meaning to his data, would tend to use habits of thought quite different from those employed by Darwin. The natural inclination would be merely to impose an oversimplified evolutionary rationalization upon the observations. The evolutionary theorist, on the other hand, would look at the facts in order to confirm the predictions of his hypotheses. Thus Dewey, in attempting to reconcile the ideas of Darwin with those of William James, assumed that all characters must be adaptive and tried to show that antithetical behavior patterns are really serviceable.[89] Therefore Dewey completely missed the point of Darwin's argument: behavioral properties may be mixtures of adaptations and historical accidents.

That biological thinking was alien to psychology, especially at the time of Darwin, may be attributed in part to the fact that psychology arose as a subdiscipline of philosophy. It is particularly revealing to note that many psychologists of considerable historical importance were interested in evolution; yet they were influenced not by Darwin, but by Spencer: Hughlings Jackson and Sechenov are examples.[90] At the time of their publication, the evolutionary pseudo-explanations of Spencer were obviously quite attractive to speculative intellects, but they proved useless as guides to scientific research. For Darwin's positive influence on psychology, one must look to indirect contributions by neurophysiologists and neuroanatomists, who were in a better position to incorporate evolutionary thinking into the structure of their theories.

There is some hope that in the future a genuine reconciliation between psychological and evolutionary thinking may be effected. The kind of model that Darwin applied is finding increasing utility in psychological research, and the similarities between the two subject matters provide a useful source of new ideas. The theory of operant conditioning, for example, has many parallels with that of natural selection. For instance, it often happens that an animal will develop a particular conditioned response, when the intent was actually to condition for some other response, simply because the animal is engaged in an extraneous activity when reinforcement occurs.[91] Skinner has called this sort of accidental conditioning "superstition." It may readily be compared to such evolutionary phenomena as sexual selection and the results of pleiotropy. Just as in natural selection what matters is not adaptation but reproductive success, so in operant conditioning the fundamental relationship is not the task set by the experimenter, but the correlation between reinforcement and what the animal happens to be doing at the time. The result in

both cases is the origin of maladaptive or nonadaptive characteristics. And in both biology and psychology, the crucial methodological innovation, one which Darwin understood very well, has
been to cease trying to rationalize appearances as owing to reason
or design, and to begin asking meaningful questions about the
processes that might have generated the observed phenomena.
With a deepened understanding of theoretical ideas, a new field
for interdisciplinary synthesis could emerge—not unlike that
which Darwin achieved for himself.

Turning to the ethologists, or zoologically orientated students
of animal behavior, one has little difficulty in seeing why they
have failed to give Darwin the credit he deserves. Instead of
going back to fundamentals, as did the founder of the science,
and constructing theoretical models to be tested by means of the
hypothetico-deductive approach, they have largely relied on
more primitive forms of induction. What theoretical systems
they have elaborated contain little reference to evolutionary
theory.[92] In conformity with the European tradition, they have
tended to employ the sort of typological comparison that still
prevails among many morphologists. They have simply gathered
facts, put similar behavior patterns together, and superimposed a
historical rationalization. Thus Lorenz states: "It is an inviolable
law of inductive natural science that it has to *begin* with pure
observation, totally devoid of any preconceived theory and even
working hypothesis." [93] A more pernicious fallacy could
scarcely be enunciated. Darwin, in all of his work, including that
on behavior, proceeded with a diametrically opposite methodological assumption. Small wonder that he has not received the
recognition he deserves.

We might reemphasize the fact that anthropologists have lately
come to appreciate Darwin's *Expression of the Emotions* as being
full of useful ideas for the study of human nonverbal communication.[94] This is understandable, in view of the great utility of
comparative methodology in anthropological research. There was
little justification when Freud extrapolated to all mankind the
results of a study on a few Victorian neurotics. And it is easy to
see why many of the soundest criticisms of psychoanalysis have
come from such anthropologists as Malinowsky. Generalizations
about organisms, including man, should be tested by showing
how well they stand up to experience in the widest variety of

circumstances. The diversity of organisms and cultures in nature gives a priceless opportunity for criticizing and developing theoretical ideas. That Darwin took full advantage of such opportunities is one of the major reasons for his success.

9. *Sexual Selection*

Of all Darwin's works, it is *The Descent of Man* that in many ways best manifests the nature of his thought. Here certain major theses, elsewhere developed separately, are drawn together into an intricate, yet unitary, system of ideas. But to understand the work is no mere problem of seeing the obvious, for it may be read upon more than one level, and it builds upon concepts which are fully developed only elsewhere in Darwin's writings. Even the title is misleading. The subject is not man, but sex: *The Descent of Man, and Selection in Relation to Sex.* Darwin as usual organized his argument around a vast body of empirical facts which, in a sense, could speak for themselves. Yet the arguments are all too readily passed by, and the facts too easily seen, not in relation to the theories which they support, but as raw data, to be taken "as they are." The hostile reader easily ignores the basic premises, however well conjoined they are to lend each other mutual support, or however artfully the conclusions are derived from the premises. And with all the intricacies of logic to consider, the argument requires the closest attention if it is to be followed at all.

In *The Descent of Man* the major theme is *sexual selection*, a topic which Darwin could develop only in bare outline in *The Origin of Species*, where he gives the following definition:

> This depends, not on a struggle for existence, but on a struggle between the males for possession of the females; the result is not death to the unsuccessful competitor, but few or no offspring. Sexual selection is, therefore, less rigorous than natural selection. Generally, the most vigorous males, those which are best fitted for their places in nature, will leave most progeny. But in many cases, victory

will depend not on general vigour, but on having special
weapons, confined to the male sex.[1]

And, especially among birds, a major struggle seemed to occur
when the males competed in their efforts to woo the females
during courtship. This theory could be invoked to explain a great
deal, and Darwin generalized that "when the males and females of
any animal have the same general habits of life, but differ in
structure, colour, or ornament, such differences have been mainly
caused by sexual selection; that is, individual males have had, in
successive generations, some slight advantage over other males, in
their weapons, means of defence, or charms; and have transmitted these advantages to their male offspring." [2] As sexual selection casts light upon various facts which seem inexplicable by
means of natural selection, Darwin's critics have charged that
sexual selection is an *ad hoc* hypothesis, designed to explain away
difficulties in his theory.[3] And even some of Darwin's most enthusiastic supporters have overlooked the strategic importance of
the concept.[4] In fact, sexual selection is Darwin's most brilliant
argument in favor of natural selection, of which it is a corollary.
Selection, whether artificial, natural, or sexual, depends upon
differential reproductive success, and not on adaptive advantage
to the individual or group. If, owing to some historical accident,
it happened that an organism had some characteristic that was
otherwise useless or even deleterious, but which did enable him
to leave more progeny than other members of his sex, then perhaps that character would become more abundant in the population. Thus, a greater ability to attract the opposite sex could
evolve in spite of its being unnecessary for procreation. Seeing
evidence for such phenomena in the ornaments and excessive
bodily size of the male in various mammals, as well as in the
brilliant plumage of birds, Darwin realized that the occurrence of
even one such structure was fatal to the hypothesis of design. As
in orchids, evolution produces contraptions, not contrivances.
And of course, the notion of emergent evolution, which still
attracts some uninformed following, is likewise demolished by it.
The entire argument on sexual selection, which occupies the
greater portion of *The Descent of Man*, thus provides a critical
test not only of natural selection, but also of the various contrary
hypotheses. It is an argument which the critics of Darwinism
have never really faced, except to assert that it is an *ad hoc*

hypothesis, and to elaborate some *ad hoc* hypotheses of their own.

As to the factual issue of whether or not sexual selection has occurred, there is no reasonable doubt that at least some evolutionary changes must be attributed to it. Mathematical genetics can predict the conditions under which it should occur.[5] And, leaving aside the few who contest the issue because to do otherwise would amount to rejecting their own points of view, biologists competent to judge agree that the facts bear out the predictions from theory.[6] It is only on details that there are grounds for disagreement, and the assertions to the contrary so often encountered in works on the history or philosophy of science are wrong.[7] But the whole problem is very intricate, as will be clear from the analysis to follow. The difficulties involved may serve as warning to those who casually pass judgment either on the problem or on its greatest theorist.

CONCEPTUAL BASES OF SEXUAL SELECTION

We may begin the analysis with some important distinctions. The first is the distinction between sexual selection and natural selection for the functional improvement of the reproductive apparatus.[8] To Darwin, sexual selection "depends on the advantage which certain individuals have over others of the same sex and species solely in respect of reproduction." [9] Mere differences in ability to reproduce, such as greater fertility or more effective copulation, are subject to natural selection. Only where there is no advantage, say, in competing with other species, can selection be purely sexual; any competition would have to be between members of the same species. Clearly, a trait may have both advantages. The distinction is not absolute, and Darwin fully realized this.[10] In order to establish his thesis, he had to distinguish between the effects of natural and sexual selection. Therefore his examples are carefully chosen to show that in certain instances natural selection is not sufficient to account for the properties in question. We may draw upon an example using horns, structures which could be used either to gain advantage in sexual combat or to ward off a predator. In antelopes, some species have long, curved horns, which project over the animal's back and seem poorly adapted for fighting. Darwin remarks:

. . . when two of these animals prepare for battle, they kneel down, with their heads between their forelegs, and in this attitude the horns stand nearly parallel and close to the ground, with the points directed forwards and a little upwards. The combatants then gradually approach each other, and each endeavours to get the upturned points under the body of the other; if one succeeds in doing this, he suddenly springs up, throwing up his head at the same time, and can thus wound or perhaps even transfix his antagonist. Both animals always kneel down, so as to guard as far as possible against this manoeuvre. It has been recorded that one of these antelopes has used his horns with effect even against a lion; yet from being forced to place his head between the forelegs in order to bring the points of the horn forward, he would generally be under a great disadvantage when attacked by any other animal. It is, therefore, not probable that the horns have been modified into their present great length and peculiar position, as a protection against beasts of prey.[11]

It was possible, then, to construct an argument which would eliminate some of the variables. In the example of horns, the possibility of natural selection was carefully ruled out. And as we shall see, the evidence was weighed to eliminate such other complications as the effects of pleiotropy and ecological differences between the sexes. With a well-designed theoretical system, it was possible to predict where the effects of different causes would be most pronounced, and to refer to factual data for verification.

The other important distinction is that between male combat and female choice. On the one hand, there is the fighting between males, with the accompanying structures that give advantage in such combat: antlers, great physical strength, or the spurs of fowl. On the other hand, we have the curious ornaments of peacocks and birds of paradise, which Darwin thought had developed because the females were more attracted to the males of striking appearance. This second version of sexual selection has been more controversial than male combat. The reason why female choice is accepted or not reflects to some degree the difficulty of the problem and the ambiguity of the evidence, but there also exist basic disagreements about theoretical premises, methodology, and even metaphysics. Yet it should be noted here and now that rejecting female choice does not force one to repudiate sexual selection; and male combat is an equally devastating argument against design. Further, it may help at this point to refute the charge that Darwin attributed a sense of beauty, in an anthropomorphic sense, to fowl.[12] He simply maintained that animals may, in some instances, be differentially attracted or repelled by various patterns or colors. His position is stated as follows: "When, however, it is said that the lower animals have a sense of beauty, it must not be supposed that such sense is comparable with that of a cultivated man, with his multiform and complex associated ideas. A more just comparison would be between the taste for the beautiful in animals, and that in the lowest savages, who admire and deck themselves with any brilliant, glittering, or curious object." [13] Whether one wishes to identify such phenomena with human tastes, and how closely the analogy holds, was utterly irrelevant to Darwin's argument. There is no scientific reason for supposing that a basic difference exists; but naturally, human vanity and arrogance are repelled by such comparison.

ROOTS OF THE HYPOTHESIS IN OTHER THEORIES

Further insight into how sexual selection fits in with evolutionary theory in general emerges when we consider the historical roots of the hypothesis. It has been maintained that he borrowed the idea from his grandfather Erasmus Darwin.[14] But the evidence, such as it is, is hardly compelling. It is true that Erasmus Darwin entertained various notions concerning the evolutionary significance of sex, but these must not be confused with sexual selection.[15] The idea of sexual vigor being advantageous to the species, or of its constituting a driving force in causing evolutionary progression, could be translated into natural, but not sexual, selection, but only if the population model were understood. Perhaps the elder Darwin's writings, with their great emphasis on sex, were of some heuristic utility. On the other hand, the structure of Darwin's argument suggests another source. Sexual selection has much in common with artificial selection and can almost be treated as a variant of it. The nexus is quite obvious in *The Descent of Man:*

Just as man can improve the breed of his game-cocks by the selection of those birds which are victorious in the cockpit, so it appears that the strongest and most vigorous males, or those provided with the best weapons, have prevailed under nature, and have led to the improvement of the natural breed or species. A slight degree of variability leading to some advantage, however slight, in reiterated deadly contests would suffice for the work of sexual selection; and it is certain that secondary sexual characters are eminently variable. Just as man can give beauty, according to his standard of taste, to his male poultry, or more strictly can modify the beauty originally acquired by the parent species, can give to the Sebright bantam a new and elegant plumage, an erect and peculiar carriage—so it appears that female birds in a state of nature, have by a long selection of the more attractive males, added to their beauty or other attractive qualities.[16]

The analogy is so clear-cut—showing as it does parallels with both male combat and female choice, while the conceptual transition necessary to pass from one idea to the other appears so small and natural—that the connection seems not at all improbable. How to verify it is another matter. In the available portions of Darwin's notebooks on transmutation of species, the discussion of sexual dimorphism concentrates on the morphogenetic aspects of the problem.[17] Sexual selection is mentioned in the "Sketch" of 1842, but it is virtually passed over in favor of selection for other kinds of reproductive fitness.[18] In the "Essay" of 1844, one may note the same de-emphasis, with sexual struggle promoting racial vigor, yet with the added point that adaptively deleterious characters might arise.[19] In the joint paper by Darwin and Wallace, the latter does not mention sexual selection, while Darwin devotes a disproportionately large section to it, which suggests that he thought it sufficiently important to include in a preliminary note.[20] *The Origin of Species* gives prominence to sexual selection, but there is no attempt to verify it.[21] Yet when one apprehends how much of an intellectual tour de force was Darwin's ultimate argument, it is easy to see why he postponed coming to grips with the problem. The subject is barely mentioned in *The Variation of Animals and Plants Under Domestication,* and it is clear from Darwin's correspondence that his theoretical system was largely developed after he had finished with artificial selection.[22] But about all that one can infer from these sequences and correlations is that the theoretical elaboration and verification of sexual selection drew strongly upon the study of artificial selection and embryology—as can be established from the structure of the argument.

TESTING THE HYPOTHESIS: BEHAVIOR

Verification of the hypothesis was largely based upon behavior. Darwin reasoned that a certain level of neurophysiological complexity was necessary before an organism could exercise what may be called choice. Sexual selection should therefore correlate with the degree of development of the central nervous system and sensory organs. He was able to show that sexual ornamentation is most developed among the more advanced organisms, such as arthropods and vertebrates, while it is lacking in such forms as worms and corals.[23] The difficulty of seeing through the mass of facts necessitated a great deal of ingenuity, for brilliant coloration in certain invertebrates was apt to be misleading. It had to be shown that in some instances other causes were involved: as when sea slugs are camouflaged on brilliant sponges, or their color "warns" predators that they are distasteful.[24] Occasionally Darwin found that sexual selection did not appear to have occurred where it was expected on the basis of theoretical considerations. For example, he writes: "Even in the highest class of the Mollusca, the Cephalopoda or cuttlefishes, in which the sexes are separate, secondary sexual characters of the present kind do not, as far as I can discover, occur. This is a surprising circumstance, as these animals possess highly-developed sense-organs and have considerable mental powers, as will be admitted by every one who has watched their artful endeavours to escape from an enemy." [25] Now that cephalopods, especially octopi, are favorite subjects for behavioral studies, it is known that the kind of property which Darwin expected does in fact occur.[26] But an absence of sexual selection in this group would have been purely negative evidence.

We may observe many applications of behavior in critical tests of hypotheses in Darwin's discussions of brilliant coloration. He explains his methodology as follows:

> It should be borne in mind that in no case have we sufficient evidence that colours have been thus acquired, except where one sex is much more brilliantly or conspicuously coloured than the other, and where there is no difference in habits between the sexes sufficient to account for their different colours. But the evidence is rendered as complete as it can ever be, only when the more ornamented individuals, almost always the males, voluntarily display their attractions before the other sex; for we cannot believe that such display is useless, and if it be advantageous, sexual selection will almost inevitably follow.[27]

To put this in somewhat more philosophical terms, contrary explanations could be refuted by showing that the behavior is explicable only in terms of a sexual relationship. Most of Darwin's argument, therefore, is devoted to showing that the behavioral properties are intricately adjusted so as to maximize the effects of sexual stimulation. The argument is basically the same as that which he used to verify his hypotheses about the functioning of orchids.

The behavioral mechanisms which play a role in reproduction act in a manner allowing many other testable predictions from theory. For instance, in order for there to be sexual selection by female choice, the animals must be able to perceive differences in their prospective mates. It follows that in any group of animals, sexual ornamentation will be of such a nature that the animals can perceive it. To test the hypothesis that a trait has arisen by sexual selection, one discovers which properties the animals can perceive, and sees if it is in these that the sexes differ. Thus, brilliant coloration in blind animals must be due to something other than sexual selection. This same kind of argument was used to verify the hypothesis of artificial selection; Darwin showed that the differences between domesticated forms are almost exclusively in externally visible characteristics.[28] And he carefully enumerates one group of domesticated organisms after another in which the major differences from wild relatives are just those properties which man values most.[29]

Darwin's method of testing his hypothesis by reference to be-

havior and neurophysiology is basic to a thorough survey of
sexual dimorphism throughout the animal kingdom.[30] Even
where the tests are quite straightforward and simple, they are
marked by a characteristic ingenuity and elegance. For instance:

> The bright colors of many butterflies and of some moths
> are specially arranged for display, so that they may read-
> ily be seen. During the night colours are not visible, and
> there can be no doubt that the nocturnal moths, taken as
> a body, are much less gaily decorated than butterfiies, all
> of which are diurnal in their habits. But the moths of cer-
> tain families . . . fly about during the day or early eve-
> ning, and many of these are extremely beautiful, being far
> brighter coloured than the strictly nocturnal kinds.[31]

This argument supports a relationship between visibility and
coloration. Whether this had anything to do with sex had to be
determined, of course, by means of other evidence.

For an equally delightful example, consider Darwin's treatment
of some curious ornaments on the wing-feathers of the Argus
pheasant. In this species the wings are provided with spots, which
look rather like a ball and socket. He begins with a comparison of
analogous structures in other organisms, then traces the phylogenetic and embryological development of the pattern, establishing the manner of its gradual, orderly formation in conformity
with the concepts of developmental mechanics.[32] Then he considers precisely how the spots would appear in the eyes of the
female and ties his explanation in with the artist's craft of providing optical illusions:

> In a photograph . . . of a specimen mounted as in the act
> of display, it may be seen that on the feathers which are
> held perpendicularly the white marks on the ocelli, rep-
> resenting light reflected from a convex surface, are at the
> upper or further end, that is, are directed upwards; and
> the bird whilst displaying himself on the ground would
> naturally be illuminated from above. But here comes the
> curious point, the outer feathers are held almost horizon-
> tally, and their ocelli ought likewise to appear as if illu-
> minated from above, and consequently the white marks
> ought to be placed on the upper sides of the ocelli; and
> wonderful as is the fact they are thus placed! Hence the
> ocelli on the several feathers, though occupying very dif-

ferent positions with respect to the light, all appear as
if illuminated from above, just as an artist would have
shaded them.[33]

This insight can scarcely have been arrived at by "Baconian"
induction. Nobody, gathering data at random, will obtain and
simultaneously relate to the theory of natural selection intricate
details about the morphogenesis of tail-spots, their appearance to
the female in certain kinds of light, and their ideal pattern in
relation to the art of painting. Theoretical reasoning, here as
throughout Darwin's work, must have played a determining role
in his success.

Darwin's system included a consideration of other biological
phenomena which might have been confused with the products
of sexual selection, such as warning coloration and mimicry. And
he employed the same basic techniques. Thus it seemed reasonable that lowly organized yet conspicuous animals must not have
become brilliantly colored through sexual selection, since their
nervous systems are too simple. Darwin reasoned that brightly
colored caterpillars ought to be distasteful, and if so, they should
wander about in broad daylight.[34] Yet forms not protected by an
unpleasant taste should be well camouflaged, and should only
emerge from concealment under cover of darkness. Similarly,
where two species are brightly colored, and one is highly distasteful, while the other has no such protective mechanism, it is
reasonable to infer that the latter is a mimic, obtaining protection
by its resemblance to the distasteful form, or model. Faced with
such evidently protective resemblances, one may draw upon a
variety of Darwinian techniques for verification. One test is to
compare the mimic, the mimic's closest relative, and the model,
predicting that the mimic will differ from its relative and resemble the model in those properties used by predators to recognize
the model. Darwin applied this sort of reasoning to mimetic
polymorphism in butterflies.[35] In certain species, only the females are mimics, while the males retain the pattern characteristic
of related forms. The genetic basis of this polymorphism has only
recently been worked out, and for some time was controversial.[36] Darwin proposed two explanations having certain analogies
to theories which were developed at a much later period. On the
one hand, he invoked an embryological cause: "The successive

variations by which the female has been modified have been transmitted to her alone."[37] Most modern geneticists would probably favor a second explanation, also proposed by Darwin, because of the relative ease with which modifier genes could control the pattern and restrict it to one sex.[38] He suggested that the development of mimicry in the male was prevented because the mimics were less attractive than normal forms to the females.[39] He went on to note that the males of certain mimetic forms retain some of the color pattern characteristic of their nonmimetic relatives; in addition, behavioral traits supported the view that the patterns are used in courtship.[40] This same explanation was given many years later by Ford—who overlooked Darwin's priority and did not elaborate the elegant test of the hypothesis.[41] The history of the problem of mimicry is in itself a worthy object of study, involving as it does basic issues in theory and the thought of many important scientists.[42] The topic was long controversial, but it is probably fair to say that all the theoretical objections to it have now been answered. Most of the counterarguments were, in fact, logically fallacious. For instance, it was purely negative evidence to reject protective coloration on the grounds of there being no "proof" that birds can distinguish between insects of different appearance—and positive evidence is now copiously available to show that they can.[43] Again, the notion that a mimic must resemble its model very closely to obtain protection at all is a mere vestige of typological thinking. By contrast, Darwin's methodology for testing hypotheses of mimicry, and for sorting out the joint effects of mimicry and various kinds of sexual and reproductive selection, was exceedingly rigorous, and it is not surprising that subsequent investigation has largely supported his views.

ADDITIONAL TESTS OF THE HYPOTHESIS

In addition to behavior, other influences would be expected to determine the conditions under which sexual selection might occur. For instance, it ought to be most frequent in those forms where there is a pronounced inequality in numbers between the sexes. In monogamous species in which the sex ratio is one-to-one, the males should all obtain mates in spite of the females' preference for certain individuals, and sexual selection should ordinarily have no effect. Conversely, in forms with a marked disparity in numbers between the sexes, the probability for sexual selection would be high. Darwin realized that in polygamous forms the effect would be the same as an actual difference in sex ratio, and that differences between the sexes would therefore be accentuated in such animals. He demonstrates, by a detailed enumeration, that the expected correlation does in fact obtain.[44] He notes, for example, that among terrestrial carnivores, only the lion is both markedly sexually dimorphic and polygamous. Similarly, seals very commonly display both polygamy and dimorphism in the secondary sexual characteristics. It has been objected—sometimes ignoring Darwin's answer—that sexual selection appears to have occurred in monogamous species.[45] Darwin constructed what amount to statistical models to show that sexual selection can occur in monogamous forms.[46] Suppose that in a given species the males were all ready to breed before the females, and that the females only begin to select mates gradually. At the beginning of the mating season the females, being few in number, would be able to select mates from many alternatives. If it were true that the most healthy and vigorous females were first to begin breeding, then the most attractive males would obtain the mates best fitted to raise offspring. Other models are possible, but they are superfluous when the logical reason for failing to think of them is understood.[47] The objection is due to typological thinking: there is "the male" which pairs with "the female" and has offspring which resemble him. But there is no "the male" or "the female." The real question is one of finding conditions that predispose some members of one or the other sex to leave disproportionate numbers of offspring.

The remarkable complexity of Darwin's thought is particularly well revealed in his discussion of sexual dimorphism in birds.[48] Here he interrelates the major themes in *The Descent of Man*—namely, behavior, the laws of variation, and natural selection—to construct one of the most sophisticated arguments in his system. In the fourth edition of *The Origin of Species*, Darwin suggested a possible explanation for the fact that where one sex is more conspicuous than the other, it is usually the male.[49] He maintained that the female is exposed to great danger when incubating the eggs, and that, therefore, selection prevents change in the direction of conspicuousness. Wallace supported this hypothesis and argued that although brilliant coloration must have tended to develop equally in both sexes, it became suppressed in the female. In his later thinking, Darwin came to reject his earlier hypothesis, although Wallace's views were scarcely altered.[50] Darwin's change in opinion resulted from his analysis of the role of developmental mechanisms in determining evolutionary events. His thinking, if a bit intricate and, as he said, "tedious," could hardly better exemplify his manner of reasoning.[51] He observed that some characters develop in only one sex, and that when thus limited, they very often develop late in life. He did not explain the phenomenon, which we would now attribute largely to the action of sex hormones, except to relate it to his ideas on the physiology of variation. He noted that certain traits limited to one sex—for example, the antlers of some deer—do not develop if the gonads are removed. Yet if, as in the antlers of reindeer, the secondary sex characters occur in both sexes, castration does not prevent their development.[52] The same kind of relationship could be seen in birds: he explained the fact that old hens often come to resemble roosters as the result of disturbances in the reproductive system.[53] He could thus distinguish two kinds of embryological mechanism capable of producing characters which might be affected by sexual selection. If we oversimplify for the sake of argument, these would be, in modern terms, those which are influenced by sex hormones and those which are not. Whether or not a property was controlled by sex hormones would be fortuitous. And under most circumstances selection for the character would primarily affect the reproductive success of only one sex. If the males happened to have a hormonally controlled character favored by selection for its presence in the male, whatever might be its effects on the female, the result would be a population sexually dimorphic for that character. If, on the other hand, the character were not hormonally controlled, yet had an advantage for either sex, both males and females would tend to evolve it. Thus, as Darwin was able to document, individuals might possess certain characters in spite of their being harmful both to themselves and the species, through the effects of correlated variability, or pleiotropy. For example, his studies on both domestic and wild fowl showed that spurs, which are deleterious in that they interfere with incubation, are not always repressed in the female.[54] If a means of inhibiting their growth were available, it would be selected—but the evidence seemed to demonstrate that the occurrence of such a variation was not inevitable. And although a partial explanation for sexual dimorphism in the coloration of birds would be that the females acquired an inconspicuous coloring solely because it concealed them while nesting, this could not account for the fact that in forms which do not incubate, such as reptiles, amphibians, fish, and butterflies, the dimorphism is nonetheless present.[55]

In a manner typical of his reasoning, Darwin applied systematics in verifying the hypothesized influence of developmental mechanisms on the evolution of sexual dimorphism. Even before he had discovered natural selection, Darwin had become aware of the fact that the coloration of birds displays a consistent pattern of variation in relation to age and sex.[56] His conception of the development of secondary sexual characteristics was a major element in the underlying system in *The Variation of Animals and Plants under Domestication*.[57] In *The Descent of Man*, he tested out these ideas by predicting that morphogenetic mechanisms, in concert with other influences (including sexual selection) would impress a certain kind of order on the evolutionary process. As the same kind of variability should tend to characterize closely related forms, and as the kind of change that has taken place could be inferred by reference to the developmental mechanism, the different causes could be abstracted by means of a systematic survey. Let us see how this was accomplished.

It may be recollected how Darwin observed that characteristics developed late in life tend to be inherited by one sex only, and that he only partially understood the reason why. However, he very carefully analyzed the relationships between the sequence of developmental events and the increasing divergence in character as organisms of different groups mature. He explained a number of phenomena by showing how variations come into being at different stages in the life cycle. Naturally, it is fortuitous at which particular stage a variation, that is to say, a modification in developmental processes, will occur; just as it is fortuitous whether characteristics are sexually limited at first appearance or not. From such considerations it follows that the characteristics of males and females at different times of the life cycle would be determined by the properties of the morphogenetic mechanisms,

and to some degree would be independent of ecological conditions and other possible causes of evolutionary change. A theory
concerning the evolution of coloration could therefore be tested
by its implications for the timing and sequence of changes in
individual development. Darwin's argument was thus an embryological one. It takes the form of a series of rules, which he sum-
marizes as follows:

> I. When the adult male is more beautiful or conspicu-
> ous than the adult female, the young of both sexes in
> their first plumage closely resemble the adult female, as
> with the common fowl and peacock; or, as occasionally
> occurs, they resemble her much more closely than they
> do the adult male.
>
> II. When the adult female is more conspicuous than the
> adult male, as sometimes though rarely occurs, the young
> of both sexes in their first plumage resemble the adult
> male.
>
> III. When the adult male resembles the adult female,
> the young of both sexes have a peculiar first plumage of
> their own, as with the robin.
>
> IV. When the adult male resembles the adult female,
> the young of both sexes in their first plumage resemble
> the adults, as with the kingfisher, many parrots, crows,
> hedge-warblers.
>
> V. When the adults of both sexes have a distinct win-
> ter and summer plumage, whether or not the male differs
> from the female, the young resemble the adults of both
> sexes in their winter dress, or much more rarely in their
> summer dress, or they resemble the females alone. Or the
> young may have an intermediate character; or again they
> may differ greatly from the adults in both their seasonal
> plumages.
>
> VI. In some few cases the young in their first plumage
> differ from each other according to sex, the young males
> resembling more or less closely the adult males, and the
> young females more or less closely the adult females.[58]

One may see by inspection that although these rules encompass
that which occurs in nature, they by no means exhaust the possibilities. For instance, there is no group in which adult males and
both male and female juveniles are brilliantly colored and the

adult females are cryptic, as would be expected if selection had acted to conceal the nesting female. By such reasoning, Darwin tested the conflicting theories, and came up with the conclusion that whether, and how, the sexes differed is largely due to the kind of inheritance responsible for the variations.

The question has been raised why Darwin overlooked the phenomenon of mutual display, in which members of both sexes stimulate each other.[59] One suggestion has been that perhaps Darwin had an "all or nothing" view. Yet in one of those papers which nobody bothers to read, he discusses analogous phenomena in both birds and monkeys, observing that display of ornaments used in courtship is not restricted to the breeding season.[60] He points out that in such instances one cannot tell whether the brilliant coloration originated through a purely sexual selection or not. Mutual displays in the strict sense of the term would not be expected, and do not occur, in Darwin's system because he was unaware of their significance. He did not discover them simply because he did not look for them, as the facts available to him could be coherently explained in terms of a combination of his embryological and sexual theories. That Darwin did not proceed to consider at any length such possibilities as mutual displays supports the hypothesis developed here as to the nature of his reasoning, for one would expect him to stop when he had supported his theory and had shown its capacity to withstand rigorous criticism. For further corroboration, one may compare his position with that of Wallace.[61] Wallace, who was largely ecological in his outlook, came to de-emphasize sexual selection because he was so impressed with the close adjustment of organisms to their environments. He would seem to have run into difficulties with the logic of conditional statements, and to have felt that natural selection would necessarily counteract a deleterious pleiotropic effect.[62] Darwin, on the other hand, based much of his system on embryology and expected nonadaptive or maladaptive evolution to occur. Hence the structure of his theory, combined with his rigorous manner of reasoning, accounts for his definitive position.

Darwin's observations on sexual selection in man form but a minor part of his argument, no more important than the chapter on mammals, and adding little to his basic thesis. To be sure, the sections on man support his theory, for he does show that there is

a gradation of properties occurring in man and his closest relatives. From the high degree of cerebral development in man, one
would predict that sexual selection ought to be especially marked.
The properties which man selects artificially in his domestic productions have much in common with those he values in selecting
a mate. We see the analogy brought out with full force when
Darwin gives us a vivid image of a Hottentot passing down a line
of women and choosing the one "who projects farthest *a
tergo*." [63] This is all understandable in view of Darwin's reasoning from theory to fact, rather than attempting to force an evolutionary interpretation on data accumulated at random.

Darwin's basic thesis—that in certain instances sexual selection
occurs—is well established. And the study of behavior has shown
that various birds, fish, and other organisms do in fact mate selec-
tively with sexually ornamented forms.[64] Yet it is probably fair
to say that natural selection accounts for a greater proportion of
sexual dimorphism and brilliant coloration than Darwin believed.
Many such attributes seem useful in allowing the recognition of
other members of the same species, and this has such advantages
as preventing the formation of sterile hybrids.[65] Some sexual
coloration, such as that which characterizes lizards, is advantageous in male combat, rather than in attracting the females.[66] In
birds, it appears that coloration plays an important role in stimulating the female's secretion of hormones.[67] But in this last example, it is not known to what degree sexual selection may also have
been effective. Darwin's ideas on the effects of developmental
mechanisms are also sound, although modern biology would
probably attribute more to selection and maintain that modifier
genes would readily counteract the effects of pleiotropy. Yet the
logical problems involved in this issue are sufficiently difficult
that there may turn out to be more truth in his ideas than has
been thought.

Once again we have seen how Darwin's reasoning included a
strong theoretical component. *The Descent of Man* is no product
of enumerative or "Baconian" induction, but owes its success to
the power of abstract reasoning which gave rise to it. The factual
basis from which it derives support is enormous, but it is a highly
selected body of data, and one which illustrates no brute empiri-
cisms. If someone reads the works of Darwin, expecting to find

no general idea or theoretical system, he will naturally conclude that Darwin was a mere enthusiastic naturalist and a good observer, of average intelligence, who worked hard. If, on the other hand, one examines the underlying argument, it is clear that Darwin's merits as a naturalist or observer result from his having known, on the basis of theory, where to look. When men attempt to study organisms "as they are," guided by no hypothesis, the chances are vanishingly small that they will ever see anything more than organisms "as they are," and they probably will see even less.

10. Conclusions

Nothing hinders the work of genius more than man's incapacity to appreciate original ideas. The fact that Darwin's accomplishment has been controversial merely reflects the degree to which it has been misunderstood. Was Darwin just a naturalist and a good observer, or was he a theoretician of the first rank? Was he unphilosophical, or did he possess sufficient wisdom not to embrace the metaphysical follies of his detractors? Is *The Origin of Species*, in the words of a present-day historian, "the accumulated observations in a naturalist's commonplace-book," or is it one of the most sophisticated intellectual edifices that the human mind has ever devised?[1] Did Darwin heap fact upon fact, or was it fact upon theory? Do his interests suggest narrowness of understanding, or has the scope of his synthesis proved too much for his critics? The answers are to be found in his works. To learn of the facts, one reads the latest journals. To understand biology, one reads Darwin.

Yet one still may ask just why it was that Darwin attained such remarkable success. For the mere fact of his having produced a series of classic works leaves out the personal element of why it was he, rather than someone else, who wrote them. Intellect alone does not explain why he synthesized upon such vast scale. Zeal or labor are far too common, and his accomplishment was too far beyond the ordinary, to provide a satisfactory answer. The notion that he stole his ideas from his predecessors fails completely with many of his more original ideas. Yet Darwin's success may readily be explained by a very simple hypothesis which seems not to have occurred to his critics: he thought. He reasoned systematically, imaginatively, and rigorously, and he criticized his own ideas. We have seen, in examining his theories one by one, that he applied much the same manner of reasoning to most varied problems. His models were rich in implications, and he knew full well how to exploit a promising lead. Through his grasp of the logic of his argument, he saw what facts to seek for critical testing of his hypotheses. It is not difficult to imagine how the development of one system of ideas could generate yet another hypothesis: as when mountain building gave rise to the coral reef theory, artificial selection led to natural and sexual selection, and fertilization mechanisms in orchids gave a hint for heterostyly and the prevalence of outcrossing. His way of thinking—in effect, his heuristic —provides that explanation which alone can withstand rational criticism.

The unity of thought underlying his work as a whole is manifested by the formal parallels between Darwin's various theories. Several involve selective-retention models: the coral reef hypothesis, natural selection, and some ideas on climbing plants. Others invoke geographical patterns in verification: coral reefs, and several arguments for evolution. We may see further parallels in the behavioral verification of mimicry and the application of the same test for selection, both artificial and sexual. Likewise, there is a distinct similarity between correlated growth and the principle of the direct action of the nervous system. Such close parallels are difficult to explain as owing to chance alone. And when one reflects upon how useful it is, in solving a scientific problem, to borrow models from other fields, one may readily apprehend how skill in doing so might make it far easier to invent significant new ideas. This is not to say that Darwin must have been conscious of any such transfer. In scientific reasoning, as in chess or painting, one may develop competence either by learning heuristic principles or by repeating successful actions. But a grasp of problem-solving techniques is clearly to a scientist's advantage, and a conscious awareness of them all the better. When it is seen how much more readily certain problems in the logic of the historical sciences are solved when considered in terms of geological examples rather than zoological ones, the utility of analogical reasoning is scarcely to be denied. Problems of morphogenesis and of psychophysical isomorphism may well be solved through the application of the same kind of model. The similarities between certain aspects of thermodynamics and selection theory are more than just a curiosity. Selective-retention models show promise for both embryology and psychology. And the nature-nurture controversy has an ecological analogue in the dispute over the relative importance of density-dependent and density-independent factors. Such generalists as Darwin have many opportunities to innovate where specialist resources prove inadequate.

In observing that Darwin's major innovations had roots in his earlier work, one might ask if this should detract from our esteem of him. One could argue that his discoveries were the mere accidental consequence of his having worked on problems which predisposed him to innovate. It is true that if Darwin had not approached the various problems from the right direction, he probably would not have attained such success. Yet we tend to forget that knowing where to seek the truth is a great talent in itself. Had this account been written to give the impression that Darwin snatched his ideas out of nowhere, he would fit the stereotype of inspired genius. But the real question to be asked is whether the stereotype is valid for anyone. We are impressed by what appear to be great leaps of the intellect. But in all probability these saltations are delusory, the artifacts of our ignorance of intermediate steps, analogous to gaps in the fossil record. When reliable data are available, they favor an orderly progression.[2] That Darwin had to proceed by gradual steps in no way detracts from the fact that he alone made them. It is one thing to stumble upon natural selection, or to use something like it to argue against evolution; it is quite another to see its significance. Darwin's capacity for developing his ideas, and for thereby generating new ones, was no common talent, and the fact that he exploited this ability so often refutes attempts to dismiss him as simply fortunate.

This generation of new ideas from old ones was a synthetic and innovative process, not to be confused with mere compiling. However true it may be that the spadework of fact-gathering is indispensable in the erection of theoretical edifices, it is rightly the architect whom we most admire. Just as boards and stones are not buildings, raw data are by no means properly called knowledge. We have an understanding only when the data are arranged in a system, and not every kind of order will do. To berate Darwin for borrowing elements in his theories, or to dismiss him as a compiler, is to misconstrue the role of theory in science. It is to confound scholarship with pedantry.

Darwin's theories were certainly original. But his work is not valued simply because of its novelty as such. He had a peculiar gift for developing ideas of the kind that scientists can use. *The Origin of Species* is less to be valued for the answers it gives than for the questions it asks. A biogeographer would naturally seize upon an evolutionary model in an effort to make sense out of his data. Paleontologists could readily see in Darwin's view, whether true or not, that the fossil record is most incomplete, a momentous issue which their personal efforts could resolve. After reading Darwin's book on orchids, no real biologist can look upon a flower as before. The evolutionary writings of Lamarck have proved barren from the time of their publication; Darwin's have continued to serve as the foundation for new research. This was Darwin's fundamental contribution to scientific thought, and it will endure regardless of whether his particular hypotheses are ultimately confirmed. This point is basic, for otherwise we cannot help but view history from a naïve and philosophically erroneous point of view. The notion that the growth of scientific knowledge has been cumulative, or that ultimate material truth is the sole criterion for the worth of a hypothesis will inevitably preclude our appreciating a system like Darwin's. It is only through understanding what such systems do for working scientists that one may really appreciate the significance of his accomplishment. There has long been a tendency, especially among philosophical physicists, to regard prediction as the basic end of scientific research. The pursuit of such omniscience as men ordinarily attribute to God has profound appeal to human ambition. And science provides man with a control over nature which likewise evokes a metaphysical pathos. Beyond the mundane possibilities of controlling his destiny, man looks to science to satisfy his metaphysical appetites. And he strives to exercise his pedantic faculty: all men, by nature, desire to have others think that they know. But science has more important, intrinsic criteria of worth. In seeking a scientific explanation, Darwin was pursuing a very special kind of understanding. Prediction was of course important, but that was not the exclusive goal. As Darwin puts it: "The only advantage of discovering laws is to foretell what will happen & to see bearing of scattered facts." [3] A theory is valued because it provides a clue as to where to look for discoveries that have yet to be made; it has a heuristic function. Viewed from without, science appears to be a body of answers; from within, it is a way of asking questions. For this reason, the crudest approximation, if it provides hints for the solution of a broad range of problems, has every advantage over the most elegant mathematical law which asserts nothing of interest. A theory, indeed, which suggests that certain occurrences cannot be predicted, is of the utmost value if it guides research into productive channels. The "predictionist thesis" and "H.-D." model seem a bit trivial as clues to what real scientists are trying to do. We have every reason to think that Darwin advocated the evolutionary classification system not because one who memorizes it can regurgitate the maximum number of brute facts, but because it facilitates the discovery of important new biological principles. Similarly, turning away from idealistic morphology to search for a reconstruction of the history of organisms is justified because the new (if more difficult) questions are fundamental to such allied sciences as genetics, embryology, and biogeography. The truly reliable guide to the importance of a theory is its utility in the dynamics of investigation, and not the emotional appeal of the finished product.

Darwin himself admitted, in characteristically modest terms, to his great originality.[4] But he did not believe that he possessed intelligence to a remarkable degree.[5] Certainly his intellectual powers, although marked by personal weaknesses and strengths, were far above average. He did experience difficulty with algebra and foreign languages, and he showed no astonishing facility for verbal reasoning. Nonetheless, he does remark upon his delight in Euclid.[6] His talent for geometry is manifest in his geology, taxonomy, embryology, and morphology. Thus it is evident that he did possess uncommon ability in certain kinds of reasoning. By the conventional indices, his intelligence quotient would probably indicate intellectual superiority, but not genius. Yet such standards can have little meaning in judging a unique individual with such unusual talents. Whatever may have been Darwin's intellectual resources, he used them with almost superhuman effectiveness. The fact of his accomplishment may be explained in at least two ways. Perhaps he was gifted with enormous power of intellect, and if so, he was indeed a remarkable man. Alternatively, he somehow accomplished all his feats of reasoning in spite of his limitations; if this latter possibility be accepted, we have reason to esteem him all the more. And there may be a residuum of truth in this second point of view. Perhaps we should attribute his accomplishment less to intelligence than to wisdom. As we have seen, Darwin was remarkable for his capacity to apply a characteristic manner of thought. His reasoning was at once imaginative and critical. While willing to entertain any hypothesis, he subjected each one to the most demanding tests. These two fundamental steps of scientific thought—the conjecture and refutation of Popper—have little place in the usual conception of intelligence. If something is to be dismissed as inadequate, it is surely not Darwin.

Darwin's works manifest the activity of a mind seeking for wisdom, a value which conventional philosophy has largely abandoned. Indeed, his science is inconceivable without his philosophy. In its formal structure, scientific thought resembles both learning and natural selection.[7] The organism adopts the successful behavior pattern, the species accumulates variations which allow it to survive, and science retains the hypotheses which withstand testing. But here the parallel ends. Learning and evolution proceed blindly, frequently leading to "superstition" or to parasitic degeneration and ultimate extinction. Yet man criticizes his ideas; and beyond that he investigates the processes which give rise to them. When his ideas lead him astray, he asks if there was something wrong with his methods. He realizes that a correlation may not indicate a cause and effect relationship, or discovers that his reasoning has been fallacious. In other words, he philosophizes. Darwin was very successful in philosophy, in the sense of attempting to gain wisdom through reflecting upon his experience, and was very careful to learn from his own mistakes.[8] Darwin's philosophical thought, which permeates his writings, was mostly of a critical nature, and professional philosophers tend to look upon such mundane reflections with scorn. Yet they have had no small significance in the history of man's dealing with nature. It may not appear particularly subtle when Darwin observes that "He who understands [a] baboon would do more toward metaphysics than Locke."[9] Nonetheless, it is the kind of idea which has most significance for human life.

The mind not attuned to technicalities is hardly likely to appreciate Darwin's real merits. His manner of thinking gives rise to no obvious spectacle. In military affairs, the public habitually applauds the charge of cavalry and ignores the feats of logistics that may be far more significant in achieving victory. Positional chess seems far less interesting, to the untrained eye, than do attempts at daring combinations. To some, it seems like cheating when a battle is won on the assembly line, or a game of chess through mastery of the openings. Likewise in science, an ability to put the ideas of others into the form of a new theory is all too frequently dismissed as mere compilation or even plagiarism. When ideas are seen to have been derived from the implications of earlier work, there seems to be less originality. Strategic maneuvers, such as shifting from genetics to embryology in order to understand variation, are rarely appreciated. Yet, in the final analysis, the real criterion of greatness in such matters is success. And when one does comprehend the subtleties of chess-playing or of scientific research, one sees much that is admirable in less conspicuous actions. Perhaps Darwin will always have most appeal to the connoisseur. But one need not understand all the fine points of an art to grasp the general concept of a masterpiece.

Continuing the foregoing analogy, we may say that Darwin's critics have both misconstrued his tactics and ignored his strategy. In noting that sometimes Darwin has made mistakes, one can easily overlook the fact that the way to correction has largely been implicit in his work. The growth of scientific knowledge has involved a continued reexamination of basic ideas, and only those theories which can be tested have ultimately proved valuable. In this process, one must consider all the possibilities, and coming up with a false, but testable, hypothesis, is no mean contribution. All scientific theories—not just pangenesis—are provisional. The success or failure of a scientist largely depends upon his methodology, which perhaps furnishes the best index of his greatness. Darwin's ability to apply his methods, time and again, to the discovery of basic principles, should be the true basis of his reputation. But many commentators on his works have overlooked the methodological explanation for Darwin's success; and their oversights are largely the result of their own defective techniques. How Darwin obtained his ideas, how he developed them, and why he advocated them are all matters of empirical fact, and can be ascertained by means of the scientific method. History is no exception to the general rule that the same kind of reasoning may be used to verify hypotheses in all scientific disciplines. There is much in common between excellent work in stratigraphic geology, comparative anatomy, and intellectual history; and the same kinds of mistakes occur in all. It would be folly to regard any treatment of Darwin's works as a body of final answers. Nonetheless, it is clear that when one approaches Darwin's ideas armed with a well-developed hypothesis, one knows when his predictions have failed. And the reliability of an argument may be evaluated by seeing if the canons of evidence have been met. Perhaps it will help to point out just where Darwin's critics have gone astray, for the history of science is itself a new field, sorely in need of a better methodology.

Good scientific investigations employ critical tests of hypotheses by serious attempts to refute them. They do not involve simply amassing data consistent with a particular interpretation, oblivious to whether or not the facts are equally consistent with another hypothesis. Darwin's critics may cull the data for as long as they wish, searching for facts suggestive of plagiarism. But what facts have been given that cannot be interpreted otherwise, unless the accusers' basic theses are true? There is no scientific argument at all when one of them seizes upon the absence of a few documents which might perhaps incriminate Darwin, to support his own point of view.[10] The imaginative literature asserting that Darwin stole his basic ideas contains little more than negative evidence and *ad hoc* hypotheses. It was completely unnecessary to reply, in essence, that Darwin cheated unconsciously.[11]

Like many procedures, the use of purely formal properties has a number of peculiar difficulties. Just as in comparative anatomy mere resemblance is not conclusive evidence for an evolutionary relationship, so analogies in the structure of theories do not contradict their having fundamental differences. Further support is essential. We have seen the dangers at issue in the confusion between natural and group selection. Another example may be drawn from the principle of the correlation of parts, which is not to be confused with Darwin's idea of correlated variability. The notion that Nature, to enlarge one structure, must diminish another occurs, among other places, in the writings of Aristotle, Goethe, and Darwin.[12] But their applications of the concept are by no means the same. To Aristotle, explaining the correlation, in deer and other animals, between the presence of horns and the absence of incisors, according to this principle, was like explaining the lack of external ears in seals as owing to the animal's deformity, a notion which really explains nothing.[13] To Goethe, the correlation between bodily elongation and the absence of limbs in snakes was a manifestation of his Platonism: the Idea is harmoniously expressed when embodied in an organism. For Darwin, such correlations were of little importance: a limited amount of food is available, and it is channeled into various organs in proportion to expediency in promoting the survival of the species.[14] The question of how Darwin came to think of the correlation principle is of great interest, and in finding an answer the parallelism between his thought and Aristotle's or Goethe's provides a useful clue. But Darwin was no Platonist or Aristotelian. Nor did he simply take his embryology wholesale from any one of his precursors. And to go so far as to argue that Darwin did not innovate, simply because his ideas have roots in earlier thought, is like saying that there is no difference whatever between the forelimb of a lizard and the wing of a bird.

The right way to infer the meaning of a scientific concept is to relate it to its context in the process of investigation. Scientists do something with their ideas, and their thought processes may be studied like any other natural phenomenon. To be sure, in so doing, one must painfully reconstruct theoretical notions and sequences of reasoning, treat one's speculations as hypotheses, and devise and execute critical tests. Such procedure is laborious, but profoundly rewarding. Every hypothesis ought to be considered, yet we need something more than mere affirmation when we are told that Darwin invoked the principle of correlated variation to explain away difficulties in his theory.[15] By contrast, when we realize that correlated variation was one of Darwin's fundamental instruments of analysis, we have a real insight which will stand or fall according to its ability to render a wide variety of facts intelligible.

Bacon wrote that it is "the first distemper of learning, when men study words and not matter." [16] The substance of an argument is propositions, not names. What matters is ideas, not the language in which they are expressed. To the intellectual historian, the consideration of language may constitute a valuable source of insight. But not every kind of linguistic argument withstands critical scrutiny. By wrenching a term from its rightful context, one can read into it just about any meaning one may please. Before one accuses Darwin of anthropomorphic thinking, one owes him the courtesy of examining the notions he sought to convey in his supposedly anthropomorphic statements. Likewise, it is clearly bad historiographic method to infer, as did one historian, that Darwin must have intended natural selection in referring to "My theory," before having read Malthus.[17] Revealingly enough, this same writer, in a religious tract, holds that modern biologists are hypocritical in denying teleology, because their language gives him, personally, a subjective impression of teleological intent.[18]

It is the height of folly to regard an inductive generalization as conclusively verified. The truth is not of this moment, and the certainty of today all too frequently evaporates in the light of tomorrow's experiment. The fact that a theory is accepted by all may only mean that everybody has made the same blunder. In attempting to evaluate Darwin's scientific competence, subsequent experience is a useful tool because it points the way to his oversights and mistakes. But ideas are not valuable simply because we now happen to subscribe to them. The history of science is no chronicle of a pilgrimage toward clearly envisioned union with some absolute truth. Rather, it records a journey into the unknown, with tentative goals and relative certainties. The explorer's virtues must not be confounded with those of the prophet. Darwin was a great scientist because he asked great questions. He was an influential scientist because he seized upon those problems which, at the time, could be exploited in further research. His works retain their interest for the working biologist because they continue to generate new and useful theories. His thoughts have been historically important because they illuminated the path of investigation, regardless of where that path may lead. Many have argued that since they, personally, disagree with Darwin's conclusions, we should heap contempt upon the scientist or his thinking. But could they, or anybody, have done as well in his place? Darwin, like every human being, had intellectual strengths and weaknesses which, as has been shown, may be discovered and corroborated by scientific means. But his virtues and faults, successes and failures, discoveries and oversights, cannot be evaluated apart from the concrete situations in which they occurred. Science advances because real human beings speculate, observe, doubt, and wonder. And only by understanding the process of investigation may we fully appreciate what a scientist has done.

Perhaps it is well to end on a more positive note and suggest how much the intellectual historian can gain through anthropological and psychological insight. Typical of Victorian gentlemen, Darwin affected great modesty, and gladly attributed his success mainly to such qualities as industry, curiosity, and the love of truth.[19] His friends claim to have regarded his character as above reproach.[20] Yet, in reading his autobiography and letters, one gets the impression that all his self-abasement conceals a personality hungering for self-esteem. He seems to have embraced his culture's ideal of the perfect man and striven to the depths of his soul to embody it. His unquestioning confidence in his father's wisdom, his determination to become a clergyman in spite of misgivings, his romantic attitudes toward science, his political liberalism in spite of his wealth, and his modesty at the attainment of success all fit the same pattern. The quality of his satisfaction is especially apparent when, in recounting an early geological triumph, he says: "All this shows how ambitious I was; but I think that I can say with truth that in after years, though I cared in the highest degree for the approbation of such men as Lyell and Hooker, who were my friends, I did not care much about the general public." [21] Darwin possessed an aristocrat's system of values. What counted was not reputation, but his personal sense of honor. It evidently mattered little to him that ignorant or foolish persons misinterpreted his writings. The satisfaction lay more with expressing the truth than with communicating it. If only the best minds understood him, so much the better. In a sense, his writing was marred by understatements and irony. Aesthete that he was, he preferred not to state, but to render his ideas. An inclination toward obscurity may be considered a fault. If society is to prosper, the public must be enlightened on important issues. And one may look upon Darwin's modesty as egotism carried to its logical conclusion. Nonetheless, Darwin's strength of character, his self-esteem, and his ambition are in no small measure responsible for his success. On seeing that there was evidence for evolution, in spite of what others had concluded, he had the courage and ability to seek out and to discover its mechanism. Grasping the potentialities of his discovery, he had the audacity to develop a comprehensive system of biological ideas on a scale which has scarcely been appreciated. To accomplish his great feats of intellect, Darwin needed a remarkable talent for judging the appropriate. His ideals and his value system predisposed him to act as he did. To prefer having an opinion which is true, to one which is popular, is not far removed from placing virtue before reputation. The same kind of idealism that leads to valuing the esteem of a select few manifests itself in a preference for the subtleties of theoretical insight.

Science is more than a body of knowledge or a method of inquiry. It becomes a part of culture and a way of life. In embracing science as a career, Darwin was satisfying many of his most basic impulses. The human race is fortunate that efforts to force him into conventional molds were abandoned in favor of letting him seek out his own destiny. Until he began to work on his real interests, Darwin showed little aptitude for learning. But the infusion of life into his studies transformed an undistinguished student into a great scholar. Darwin is noteworthy for having solved a peculiar kind of problem, upon which he brought to bear his own combination of talents. His was a victory for reason, for the imagination, and for devotion to ideals. It was also a triumph of an individual human being, imposing the stamp of his unique personality upon the course of history. He reasoned in his own peculiar way, and in the mind of man the universe was transformed. Whether one likes it or not, the age belongs to Darwin.

Appendix: Important Dates

February 12, 1809. Birth of Charles Darwin.

October 22, 1825. Darwin entered the university at Edinburgh.

October 15, 1827. Began study for ministry at Cambridge.

April 26, 1831. Received B.A.

December 27, 1831. Beagle sailed for South America.

Autumn, 1833. Darwin found remains of large, extinct mammals, associated with marine forms resembling the modern ones.

September 17, 1835. Landed in Galapagos Islands.

January 12, 1836. Arrival in Australia.

October 2, 1836. Beagle reached England.

July, 1837. Darwin began first notebook on transmutation of species.

September 28, 1838. He began to read Malthus.

January 29, 1839. He married Emma Wedgwood.

May, 1842. Coral reef book published; first draft of work on species.

July, 1844. Second draft of work on species.

October 1, 1846. Darwin began to study barnacles.

1854. He finished monograph of barnacles, and began full-time research on species.

June 18, 1858. Long work on species interrupted with arrival of manuscript by Wallace.

July 1, 1858. Joint paper by Darwin and Wallace read before Linnean Society.

November 24, 1859. *The Origin of Species* published.

May 15, 1862. Book on orchids published.

January 30, 1868. *The Variation of Animals and Plants under Domestication* published.

February 24, 1871. *The Descent of Man* published.

November 26, 1872. *The Expression of the Emotions* published.

245

January 2, 1875. *Insectivorous Plants* published.

September, 1875. *Climbing Plants* published.

December 5, 1876. Book on cross and self fertilization published.

July 9, 1877. Book on heterostylic plants etc. published.

November 22, 1880. *The Power of Movement in Plants* published.

October 10, 1881. Book on earthworms published.

April 19, 1882. Death of Darwin.

Notes

KEY TO ABBREVIATED TITLES
(References in this work are to the following editions)

Climbing Plants
C. Darwin. *The Movements and Habits of Climbing Plants,* 2d ed. (New York: D. Appleton, 1876).

Coral Reefs
C. Darwin. *The Structure and Distribution of Coral Reefs* (Berkeley and Los Angeles: University of California Press, 1962), a facsimile of C. Darwin, *Geological Observations on Coral Reefs, Volcanic Islands and on South America* (London: Smith, Elder, 1851).

Descent
C. Darwin. *The Descent of Man, and Selection in Relation to Sex,* 2d ed. (New York: D. Appleton, 1889).

"Essay"
C. Darwin and A. R. Wallace. *Evolution by Natural Selection* (Cambridge: University Press, 1958). Contains a foreword by G. de Beer, introduction by F. Darwin and Darwin and Wallace, "Tendency of Species to Form Varieties."

Expression
C. Darwin. *The Expression of the Emotions in Man and Animals* (New York: Philosophical Library, 1955).

Fertilization
C. Darwin. *The Effects of Cross and Self Fertilisation in the Vegetable Kingdom* (New York: D. Appleton, 1889).

First Notebook
G. de Beer. "Darwin's Notebooks on Transmutation of Species. Part I.

First Notebook (July 1837–February 1838)," *Bulletin of the British Museum (Natural History), Historical Series*, II (1960), 23–73.

Flowers
C. Darwin. *The Different Forms of Flowers on Plants of the Same Species* (New York: D. Appleton, 1877).

Fourth Notebook
G. de Beer. "Darwin's Notebooks on Transmutation of Species. Part IV. Fourth Notebook (October 1838–10 July 1839)," *Bulletin of the British Museum (Natural History), Historical Series*, II (1960), 151–183.

Life and Letters
F. Darwin. *The Life and Letters of Charles Darwin, Including an Autobiographical Chapter*, 3 vols. (London: John Murray, 1887).

Living Balanidae
C. Darwin. *A Monograph on the Sub-class Cirripedia, with Figures of all the Species. The Balanidae (or Sessile Cirripedes); the Verrucidae; etc., etc.* (London: Ray Society, 1854).

Living Lepadidae
C. Darwin. *A Monograph on the Sub-class Cirripedia, with Figures of all the Species. The Lepadidae; or, Pedunculated Cirripedes* (London: Ray Society, 1851).

More Letters
F. Darwin and A. C. Seward. *More Letters of Charles Darwin: a Record of his Work in a Series of Hitherto Unpublished Letters*, 2 vols. (London: John Murray, 1903).

Movement
C. Darwin, assisted by F. Darwin. *The Power of Movement in Plants* (New York: D. Appleton, 1881).

Naturalist's Voyage (1840 ed.)
C. Darwin. *Journal of Researches into the Geology and Natural History of the Various Countries Visited by H.M.S. Beagle, under the Command of Captain Fitzroy R.N. from 1832 to 1836* (London: Henry Colburn, 1840; usually cited as 1839).

Naturalist's Voyage (1845 ed.)
C. Darwin. *Journal of Researches into the Natural History and Ge-*

ology of the Various Countries Visited During the Voyage of H.M.S. Beagle Bound Round the World (London: John Murray, 1845).

Orchids
C. Darwin. *The Various Contrivances by Which Orchids are Fertilised by Insects*, 2d ed. (New York: D. Appleton, 1886).

Origin
C. Darwin. *On the Origin of Species: A Facsimile of the First Edition* (Cambridge: Harvard University Press, 1964).

Second Notebook
G. de Beer. "Darwin's Notebooks on Transmutation of Species. Part II. Second Notebook (February to July 1838)," *Bulletin of the British Museum (Natural History), Historical Series*, II (1960), 75–117.

"Sketch"
C. Darwin and A. R. Wallace. *Evolution by Natural Selection* (Cambridge: University Press, 1958).

South America
C. Darwin. *Geological Observations on the Volcanic Islands and Parts of South America* (New York: D. Appleton, 1876).

Third Notebook
G. de Beer. "Darwin's Notebooks on Transmutation of Species. Part III. Third Notebook (July 15th 1838–October 2nd 1838)," *Bulletin of the British Museum (National History), Historical Series*, II (1960), 119–150.

Variation
C. Darwin. *The Variation of Animals and Plants Under Domestication*, 2d ed., 2 vols. (New York: D. Appleton, 1887).

Variorum
M. Peckham. *The Origin of Species by Charles Darwin: A Variorum Text* (Philadelphia: University of Pennsylvania Press, 1959).

Volcanic Islands
C. Darwin. *Geological Observations on the Volcanic Islands and Parts of South America* (New York: D. Appleton, 1876).

Worms
C. Darwin. *The Formation of Vegetable Mould Through the Action of Worms* (London: John Murray, 1881).

INTRODUCTION: THE PROBLEM AND THE SOURCES

[1] L. Eiseley, "Charles Darwin, Edward Blyth, and the Theory of Natural Selection," *Proceedings of the American Philosophical Society,* CIII (1959), 94–158; C. D. Darlington, *Darwin's Place in History* (Oxford: Blackwell, 1960); J. H. F. Kohlbrugge, "War Darwin ein originelles Genie?" *Biologisches Centralblatt,* XXXV (1915), 93–111—to name only more recent proponents of this notion.

[2] E. Mayr, "Introduction," in C. Darwin, *On the Origin of Species: A Facsimile of the First Edition* (Cambridge: Harvard University Press, 1964), pp. vii–xxvii.

[3] G. de Beer, *Charles Darwin: Evolution by Natural Selection* (London: Thomas Nelson & Sons, 1963). Also G. de Beer, "Darwin's Notebooks on Transmutation of Species." *Third Notebook,* pp. 119–150, 142.

[4] *Life and Letters,* I, 93; II, 79, 286; III, 215.

[5] *Life and Letters,* II, 286.

[6] *Life and Letters,* III, 25.

[7] *Descent,* p. 606.

[8] K. R. Popper, *The Logic of Scientific Discovery,* 3d ed. (New York: Harper, 1965).

[9] *Life and Letters,* I, 69.

[10] *Life and Letters,* I, 55.

[11] *Third Notebook,* pp. 132, 134 ff.

[12] *Life and Letters,* I, 54, 74. Cf. R. M. Blake, C. J. Ducasse, and E. H. Madden, *Theories of Scientific Method: The Renaissance through the Nineteenth Century* (Seattle: University of Washington Press, 1960).

[13] E.g., C. Singer, *A Short History of Scientific Ideas to 1900* (Oxford: University Press, 1959), pp. 507–516; and many others.

[14] *Life and Letters.* For an unexpurgated version of the autobiography: N. Barlow, ed., *The Autobiography of Charles Darwin 1809–1882.* (New York: Harcourt, Brace, 1959).

[15] *More Letters.*

[16] W. Irvine, *Apes, Angels, and Victorians* (New York: McGraw-Hill, 1955).

[17] J. C. Greene, *The Death of Adam: Evolution and Its Impact on*

Western Thought (New York: Mentor, 1961); S. Toulmin and J. Goodfield, *The Discovery of Time* (London: Hutchinson, 1965).

[18] De Beer, *Charles Darwin* (above n. 3); G. Wichler, *Charles Darwin, the Founder of the Theory of Evolution and Natural Selection* (New York: Pergamon, 1961).

[19] Darlington, *Darwin's Place* (above, n. 1); L. Eiseley, *Darwin's Century: Evolution and the Men Who Discovered It* (New York: Anchor, 1961); G. Himmelfarb, *Darwin and the Darwinian Revolution* (New York: Doubleday, 1959).

[20] Cf. B. J. Lowenberg, "Darwin and Darwinian Studies, 1959–63," *History of Science*, IV (1965), 15–54; E. Mendelsohn, "The Biological Sciences in the Nineteenth Century: Some Problems and Sources," *History of Science*, III (1964), 39–59.

[21] E.g., Himmelfarb says, in *Darwin*, p. 177, with reference to Cuvier: "From the evidence of geological stratification, he deduced that catastrophes had been both violent and sudden—a conclusion all the more convincing because the particular evidence on which it was based, the geological formations of the Alps, is, in fact, difficult to interpret on any other assumption (as has been admitted by a reputable modern scientist, himself neither a disciple of Cuvier nor an opponent of evolution)." The scientist E. Nordenskiöld (a biologist, not a geologist) says, in *The History of Biology* (New York: Tudor, 1928), p. 338: "He does not, however, deserve any severe censure for the actual theory of these vast volcanic upheavals, with their resultant inundations; the geological material available for observation was still somewhat scanty and was, moreover, as far as French research was concerned, largely gathered from the Alps, with their greatly subverted formations, which even to this day are difficult to interpret, and which are peculiarly likely to induce a belief in violent upheavals." It should be noted that Himmelfarb is a partisan of the now discarded theory of macromutation, a biological analogue of catastrophism. For Darlington and Eiseley, cf. chap. 10, "Conclusions."

[22] N. Barlow, *Charles Darwin's Diary of the Voyage of H.M.S. "Beagle"* (New York: Macmillan, 1933).

[23] *Naturalist's Voyage* (1840 ed.), pp. 50, 212; *Naturalist's Voyage* (1845 ed.) pp. 52, 174–175.

[24] C. Darwin, ed., *The Zoology of the Voyage of H.M.S. Beagle, under the Command of Captain Fitzroy, R.N., During the Years 1832 to 1836.* 5 vols. (London: Smith, Elder, 1839–1843).

[25] For editions referred to in this work see Appendix, Key to Abbreviated Titles: *Coral Reefs; South America; Volcanic Islands.*

[26] For editions referred to in this work see Appendix, Key to Abbreviated Titles: *Living Lepadidae;* and *Living Balanidae.* See also C. Darwin, *A Monograph on the Fossil Lepadidae; or Pedunculated*

Cirripedes of Great Britain (London: Palaeontographical Society, 1851; and C. Darwin, *A Monograph on the Fossil Balanidae and Verrucidae of Great Britain* (London: Palaeontographical Society, 1854).

[27] Except where otherwise stated, references here are to the first edition (above, n. 2). For other editions see Appendix, Key to Abbreviated Titles, *Variorum.*

[28] C. Darwin and A. Wallace, "On the Tendency of Species to Form Varieties; and on the Perpetuation of Varieties and Species by Natural Means of Selection," *Journal of the Proceedings of the Linnean Society (Zoology)*, III (1859), 45–62.

[29] G. de Beer, "Darwin's Notebooks on Transmutation of Species." *First Notebook*, pp. 23–73; *Second Notebook*, pp. 75–117; *Third Notebook*, pp. 119–150; *Fourth Notebook*, pp. 151–183.

[30] See Appendix, Key to Abbreviated Titles: "Essay" and "Sketch."

[31] See Appendix, Key to Abbreviated Titles: *Variation.*

[32] For editions referred to in this work see Appendix, Key to Abbreviated Titles: *Orchids; Climbing Plants; Fertilization; Flowers; Movement.*

[33] C. Darwin, "A Posthumous Essay on Instinct," in C. G. Romanes, *Mental Evolution in Animals* (New York: D. Appleton, 1884), pp. 355–384.

[34] Here abbreviated *Expression.*

[35] Here abbreviated *Worms.*

1. GEOLOGY

[1] *Life and Letters*, III, 177.

[2] *Life and Letters*, I, 61.

[3] E. Bailey, *Charles Lyell* (London: Thomas Nelson and Sons, 1962); C. C. Gillispie, *Genesis and Geology: A Study in the Relations of Scientific Thought, Natural Theology and Social Opinion in Great Britain, 1790–1850* (Cambridge: Harvard University Press, 1951).

[4] K. R. Popper, in *The Logic of Scientific Discovery* (New York: Harper, 1965), pp. 140–145, holds that the degree of falsifiability is the real rationale behind the need for simplicity in theories.

[5] G. Himmelfarb, *Darwin and the Darwinian Revolution* (New York: Doubleday, 1959), p. 93.

[6] *Life and Letters*, III, 196.

[7] W. Coleman, "Lyell and the 'Reality' of Species: 1830–1833," *Isis*, LIII (1962), 325–338.

[8] Cf. *ibid.*; W. F. Cannon, "The Uniformitarian-catastrophist Debate," *Isis*, LI (1960), 38–55; J. W. Judd, *The Coming of Evolution:*

The Story of a Great Revolution in Science (Cambridge: University Press, 1912).

[9] A. Remane, "Morphologie als Homologienforschung," *Verhandlungen Deutschen Zoologischen Gesellschaft, 1954,* (1955), pp. 159–183.

[10] *More Letters,* II, 229; cf. *Descent,* pp. 90–91.

[11] *Volcanic Islands,* p. 86.

[12] *South America,* figs. 15–21, 23, 24, 26, 27; Darwin's fig. 18 is here reproduced as fig. 2.

[13] The effects of glaciation were then unknown, but this complication does not affect the interpretation of Darwin's reasoning.

[14] *South America,* chap. 9.

[15] Popper, *Logic of Scientific Discovery* (above, n. 4), p. 269.

[16] For a sound treatment of the difference between valid and nonsensical quantification in geology, cf. J. H. Mackin, "Rational and Empirical Methods of Investigation in Geology," *The Fabric of Geology,* ed. C. C. Albritton, Jr. (Reading, Mass., Palo Alto and London: Addison Wesley, 1963), pp. 135–163.

[17] *Coral Reefs,* figs. 5, 6, here reproduced as fig. 3.

[18] *Life and Letters,* p. 70.

[19] N. Barlow, *The Autobiography of Charles Darwin 1809–1882* (New York: Harcourt, Brace, 1959), p. 109.

[20] *Life and Letters,* I, 104.

[21] *Volcanic Islands,* p. 131.

[22] *Coral Reefs,* p. 141.

[23] But G. Himmelfarb, in *Darwin and the Darwinian Revolution,* bases her argument on two examples of Darwin's supposedly defective work: coral reefs and natural selection!

[24] H. W. Menard, "Foreword" to *Coral Reefs;* C. M. Yonge, "Darwin and Coral Reefs," in *A Century of Darwin,* ed. S. A. Barnett (Cambridge: Harvard University Press, 1959), pp. 245–266.

[25] *Coral Reefs,* pp. 39–40, 111–114.

[26] Menard, "Foreword" to *Coral Reefs.*

[27] J. Murray, "The Structure and Origin of Coral Reefs and Islands," *Nature,* XXII (1880), 351–355. Murray was an associate of Wyville Thompson, one of Darwin's most bitter opponents.

[28] *Life and Letters,* III, 183–184.

[29] J. W. Judd, "Darwin and Geology," in *Darwin and Modern Science: Essays in Commemoration of the Centenary of the Birth of Charles Darwin and of the Fiftieth Anniversary of the Publication of The Origin of Species,* ed. A. C. Seward (Cambridge: University Press, 1909), pp. 337–385.

[30] C. G. Hempel and P. Oppenheim, "The Logic of Explanation," in *Readings in the Philosophy of Science,* ed. H. Feigl and M. Brodbeck (New York: Appleton-Century-Crofts, 1953), pp. 319–352.

31 S. Toulmin, *Foresight and Understanding: An Enquiry into the Aims of Science* (Bloomington: Indiana University Press, 1961); J. Hospers, "What is Explanation?" in *Essays in Conceptual Analysis*, ed. A. Flew (London: Macmillan, 1963), pp. 94–119.

32 E. H. Madden, "The Philosophy of Science in Gestalt Theory," in *Readings in the Philosophy of Science*, ed. H. Feigl and M. Brodbeck (New York: Appleton-Century-Crofts, 1953), pp. 559–570, 562.

33 W. Dray, *Laws and Explanation in History* (Oxford: University Press, 1957).

34 K. R. Popper, *Conjectures and Refutations* (New York: Basic Books, 1962), pp. 33–37.

35 For further discussions of Darwin's geological work see papers cited above; also: A. Geikie, *Charles Darwin as Geologist* (Cambridge: University Press, 1909); K. Andrée, "Charles Darwin als Geologe," in *Hundert Jahre Evolutionsforschung, das wissenschaftliche Vermächtnis Charles Darwins*, ed. G. Heberer and F. Schwanitz (Stuttgart: Gustav Fischer, 1960), pp. 277–289.

2. BIOGEOGRAPHY AND EVOLUTION

1 W. George, *Biologist Philosopher: A Study of the Life and Writings of Alfred Russel Wallace* (London: Abelard-Schuman, 1964), discusses Wallace's contribution in considerable depth.

2 W. Thiselton-Dyer, "Geographical Distribution of Plants," *Darwin and Modern Science*, ed. A. C. Seward (Cambridge: University Press, 1909), pp. 298–318, 306.

3 S. Smith, "The Origin of the 'Origin,'" *The Advancement of Science*, XVI (1960), 391–401.

4 *More Letters*, I, 367.

5 G. de Beer, *Charles Darwin: Evolution by Natural Selection* (London: Thomas Nelson & Sons, 1963), p. 86.

6 Smith, "Origin of the 'Origin.'"

7 *Life and Letters*, I, 83.

8 G. Himmelfarb, in *Darwin and the Darwinian Revolution* (New York: Doubleday, 1959), chap. 5, suggests a few other pieces of negative evidence. For positive evidence supporting the contrary view, see N. Barlow, "Darwin's Ornithological Notes," *Bulletin of the British Museum (Natural History) Historical Series*, II (1963), 201–278.

9 F. Darwin, "Introduction," C. Darwin and A. R. Wallace, *Evolution by Natural Selection* (Cambridge: University Press, 1958).

10 N. Barlow, *Charles Darwin and the Voyage of the Beagle*, (New York: Philosophical Library, 1946), pp. 246–247; Barlow, "Darwin's Ornithological Notes," p. 262.

11 E.g., *Variation*, I, 9–11.

[12] *Origin*, p. 1.

[13] J. W. Judd, "Darwin and Geology," *Darwin and Modern Science*, ed. A. C. Seward (Cambridge: University Press, 1909), pp. 337–384.

[14] N. Barlow, *Charles Darwin's Diary of the Voyage of H.M.S. Beagle* (New York: Macmillan, 1933), p. 236.

[15] Barlow, "Darwin's Ornithological Notes," pp. 276–277.

[16] *Naturalist's Voyage* (1845 ed.), p. 237.

[17] Barlow, *Charles Darwin's Diary*, p. 236.

[18] *Ibid.*, p. 337.

[19] *Ibid.*, p. 383. This passage occurs in the 1840, but not the 1845, edition of the *Naturalist's Voyage*.

[20] *Ibid.*; also Barlow, *Charles Darwin and the Voyage; Volcanic Islands*, p. 102.

[21] *Origin*, chaps. 11, 12; *Second Notebook*.

[22] C. Lyell, *Principles of Geology: Being an Inquiry How Far the Former Changes of the Earth's Surface Are Referable to Causes Now in Operation*, 5th ed., 2 vols. (Philadelphia: James Kay, 1837), II, 9–98.

[23] *Origin*, p. 394.

[24] *Origin*, chaps. 11, 12.

[25] P. J. Darlington, "Darwin and Zoogeography," *Proceedings of the American Philosophical Society*, CIII (1959), 307–319, 315.

[26] *More Letters*, II, 12.

[27] George, *Biologist Philosopher* (above, n. 1), p. 135.

[28] E.g., C. Darwin, "On the Action of Sea-water on the Germination of Seeds," *Journal of the Proceedings of the Linnean Society (Botany)* I (1856), 130–140; *Origin*, pp. 358–360.

[29] *Origin*, pp. 360–363; C. Darwin, "Transplantation of Shells," *Nature*, XVIII (1878), 120–121; C. Darwin, "On the Dispersal of Freshwater Bivalves," *Nature*, XXV (1882), 529–530.

[30] *Origin*, p. 393.

[31] *More Letters*, II, 16–20.

[32] G. G. Simpson, *The Geography of Evolution* (Philadelphia and New York: Chilton, 1965), pp. 167–208.

[33] Cf. chap. 9, "Sexual Selection."

[34] *More Letters*, I, 407–408; dated March 19, 1845.

[35] M. R. Cohen, *Reason and Nature: An Essay on the Meaning of Scientific Method* (New York: Free Press of Glencoe, 1964), provides a readable discussion on this point.

[36] *Origin*, chap. 9.

[37] J. Challinor, "Palaeontology and Evolution," *Darwin's Biological Work, Some Aspects Reconsidered*, ed. P. R. Bell (Cambridge: University Press, 1959), pp. 50–100; N. D. Newell, "The Nature of the Fossil Record," *Proceedings of the American Philosophical Society*,

CIII (1959), 264–285; A. S. Romer, "Darwin and the Fossil Record," *A Century of Darwin*, ed. S. A. Barnett (Cambridge: Harvard University Press, 1959), pp. 130–152.

[38] T. S. Kuhn, *The Structure of Scientific Revolutions* (Chicago: University Press, 1962) (*International Encyclopedia of Unified Science*, Vol. II, No. 2).

[39] L. Agassiz, *Essay on Classification* (Cambridge: Harvard University Press, 1962), pp. 46, 49.

[40] A. O. Lovejoy, *The Great Chain of Being* (Cambridge: Harvard University Press, 1936).

[41] Kuhn, *Structure of Scientific Revolutions*.

3. NATURAL SELECTION

[1] *Life and Letters*, III, 220.

[2] C. Zirkle, "Natural Selection Before the 'Origin of Species,'" *Proceedings of the American Philosophical Society*, LXXXIV (1941), 71–123.

[3] L. Eiseley, "Charles Darwin, Edward Blyth, and the Theory of Natural Selection," *Proceedings of the American Philosophical Society*, CIII (1959), 94–158.

[4] G. de Beer, "The Origins of Darwin's Ideas on Evolution and Natural Selection," *Proceedings of the Royal Society of London* CLV (1961), 321–338.

[5] C. D. Darlington, *Darwin's Place in History* (Oxford: Blackwell, 1960).

[6] T. S. Kuhn, *The Copernican Revolution: Planetary Astronomy in the Development of Western Thought* (Cambridge: Harvard University Press, 1957).

[7] In *Life and Letters*, I, 87, Darwin says: "I tried once or twice to explain to able men what I meant by Natural Selection, but signally failed."

[8] *Life and Letters*, I, 85.

[9] T. H. Huxley, *Darwiniana* (New York: D. Appleton, 1893), pp. 286–287.

[10] *More Letters*, I, 140.

[11] *Ibid.*

[12] Quoted in C. F. A. Pantin, "Alfred Russel Wallace, F.R.S., and His Essays of 1858 and 1855," *Notes and Records of the Royal Society of London*, XIV (1959), 67–84, 72.

[13] *Life and Letters*, I, 83.

[14] De Beer, "Origins of Darwin's Ideas" (above, n. 4).

[15] S. Smith, "Origin of 'the Origin,'" *The Advancement of Science*, XVI (1960), 391–401.

[16] G. Himmelfarb, *Darwin and the Darwinian Revolution* (New York: Doubleday, 1959); L. Eiseley, *Darwin's Century: Evolution and the Men Who Discovered It* (New York: Anchor, 1961); S. L. Sobol, "Ch. Darwin's Evolutionary Conception at the Early Stages of its Formation," *Annals of Biology, Moscow Society of Natural History, Section in the History of Natural Sciences,* I (1959), 13–34 (in Russian, English summary); R. C. Stauffer, "Ecology in the Long Manuscript Version of Darwin's *Origin of Species* and Linnaeus' *Oeconomy of Nature,*" *Proceedings of the American Philosophical Society,* CIV (1960), 235–241.

[17] The importance of Malthus is stressed by A. Sandow, "Social Factors in the Origin of Darwinism," *Quarterly Review of Biology,* XIII (1938), 315–326.

[18] Eiseley, "Charles Darwin, Edward Blyth" (above, n. 3).

[19] E. Mayr, "Darwin and the Evolutionary Theory in Biology," *Evolution and Anthropology: a Centennial Appraisal,* ed. B. J. Meggers (Washington: The Anthropological Society of Washington, 1959), pp. 1–10.

[20] K. R. Popper, *Conjectures and Refutations* (New York: Basic Books, 1962), p. 19.

[21] *Ibid.,* chap. 5.

[22] Plato, *Republic,* Book 7.

[23] R. Robinson, *Definition* (Oxford: University Press, 1954).

[24] L. H. Hyman, *The Invertebrates: Protozoa through Ctenophora* (New York: McGraw-Hill, 1940), pp. 5, 44.

[25] A. O. Lovejoy, "Buffon and the Problem of Species," *Forerunners of Darwin: 1745–1859,* ed. B. Glass, O. Temkin, and W. L. Strauss (Baltimore: Johns Hopkins Press, 1959), pp. 84–113, 101.

[26] E. Mayr, "Speciation Phenomena in Birds," *American Naturalist,* LXXIV (1940), 249–278.

[27] D. L. Hull, "The Effect of Essentialism on Taxonomy—Two Thousand Years of Stasis (II)," *British Journal for the Philosophy of Science,* XVI (1965), 1–18, 3–4.

[28] M. T. Ghiselin, "On Semantic Pitfalls of Biological Adaptation," *Philosophy of Science,* XXXIII (1966), 147–153.

[29] G. G. Simpson, *Principles of Animal Taxonomy* (New York: Columbia University Press, 1961), p. 19; E. Mayr, E. G. Linsley, and R. L. Usinger, *Methods and Principles of Systematic Zoology,* 2d ed. (New York: McGraw-Hill, in press).

[30] C. Lyell, *Principles of Geology* . . . , 5th ed., 2 vols. (Philadelphia: James Kay, 1837), I, 285–486.

[31] *First Notebook,* p. 44.

[32] *Second Notebook,* p. 85.

[33] *First Notebook,* p. 46.

[34] *Second Notebook,* p. 90.

[35] G. de Beer, "Introduction," *Third Notebook;* Darlington, *Darwin's Place* (above, n. 5); Zirkle, "Natural Selection" (above, n. 2); all authors cited above, n. 16.

[36] G. C. Williams, *Adaptation and Natural Selection: A Critique of Some Current Evolutionary Thought* (Princeton: University Press, 1966).

[37] *Origin,* pp. 235–242.

[38] *More Letters,* I, 293; for comic relief, cf. Himmelfarb, *Darwin* (above, n. 16), pp. 301–302.

[39] Contrary to the indignant protest in Darlington, *Darwin's Place* (above, n. 5), p. 93.

[40] C. D. Darlington, "Purpose and Particles in the Study of Heredity," *Science, Medicine and History: Essays on the Evolution of Scientific Thought and Medical Practice Written In Honour of Charles Singer,* ed. E. Ashworth Underwood (Oxford: University Press, 1953) II, 472–481, 476.

[41] Williams, *Adaptation and Natural Selection,* pp. 146–155.

[42] T. Cowles, "Malthus, Darwin and Bagehot: A Study in the Transference of a Concept," *Isis,* XXVI (1937), 341–348; Sandow, "Social Factors" (above, n. 17); J. A. Thomson, "Darwin's Predecessors," *Darwin and Modern Science,* ed. A. C. Seward (Cambridge: University Press, 1909), pp. 3–17.

[43] L. von Bertallanfy, *Problems of Life, an Evaluation of Modern Biological Thought* (New York: John Wiley & Sons, 1962), p. 108.

[44] G. de Beer, "Introduction," *Fourth Notebook;* and elsewhere.

[45] Cf. T. Dobzhansky, *Genetics and the Origin of Species,* 3d ed. (New York: Columbia University Press, 1951), pp. 246–247.

[46] Discussed in M. T. Ghiselin, "On Psychologism in the Logic of Taxonomic Controversies," *Systematic Zoology,* XV (1966), 207–215.

[47] Popper, *Conjectures and Refutations* (above, n. 20), p. 197.

[48] *More Letters,* I, 387; another argument is based on neuter castes in social insects: *Origin,* p. 242.

[49] Von Bertallanfy, *Problems of Life,* p. 89.

[50] *Origin,* p. 201.

[51] *Third Notebook,* p. 142.

[52] M. T. Ghiselin, "On Semantic Pitfalls of Biological Adaptation" (above, n. 28); also G. Sommerhoff, *Analytical Biology* (London: Oxford University Press, 1950).

[53] A. R. Manser, "The Concept of Evolution," *Philosophy,* XL (1965), 18–34; cf. the rebuttal by A. Flew, " 'The Concept of Evolution': A Comment," *Philosophy,* XLI (1966), 70–75.

[54] The main advocate of conventionalism has been Eddington. For critiques, cf. K. R. Popper, *The Logic of Scientific Discovery,* 3d ed. (New York: Harper, 1965), pp. 78–84; M. Schlick, "Are Natural

Laws Conventions?" *Readings in the Philosophy of Science,* ed. H. Feigl and M. Brodbeck (New York: Appleton-Century-Crofts, 1953), pp. 181–188; L. S. Stebbing, *Philosophy and the Physicists* (New York: Dover, 1958).

55 This point is stressed by R. A. Fisher, *The Genetical Theory of Natural Selection* (Oxford: Clarendon Press, 1930), p. 36.

56 E. Mayr, *Systematics and the Origin of Species: From the Viewpoint of a Zoologist* (New York: Columbia University Press, 1942).

57 S. Toulmin, *Foresight and Understanding: An Enquiry into the Aims of Science* (Bloomington: Indiana University Press, 1961).

58 G. G. Simpson, "Historical Science," *The Fabric of Geology,* ed. C. C. Albritton, Jr., (Reading, Mass., Palo Alto and London: Addison-Wesley, 1963), pp. 24–48; cf. Popper, *Scientific Discovery,* p. 60.

59 Cf. Flew, "Comment"; A. G. N. Flew, "The Structure of Darwinism," *New Biology,* XXVIII (1959), 25–44.

60 E.g., "Essay," p. 231; cf. *More Letters,* I, 76, 114–115; the widespread contrary view is a myth.

61 *More Letters,* I, 114.

62 A. O. Lovejoy, *The Great Chain of Being* (Cambridge: Harvard University Press, 1936).

63 A. J. Cain, "Function and Taxonomic Importance," *Function and Taxonomic Importance,* ed. A. J. Cain, (London: Systematics Association, 1959), pp. 5–19, 17.

64 Williams, *Adaptation and Natural Selection* (above, n. 36).

65 Sexual dimorphism is treated in chap. 9, "Sexual Selection."

66 Cf. *More Letters,* I, 126, where Darwin compares natural selection to "denudation" in geology.

67 L. C. Cole, "Sketches of General and Comparative Demography," *Cold Springs Harbor Symposia on Quantitative Biology,* XXII (1957), 1–16.

68 Himmelfarb, *Darwin* (above, n. 16); Eiseley, *Darwin's Century* (above, n. 16), pp. 180, 184.

69 *Life and Letters,* I, 84.

70 Sobol, "Ch. Darwin's Evolutionary Conception" (above, n. 16), p. 34.

71 N. R. Hanson, in *Patterns of Discovery, an Inquiry into the Conceptual Foundations of Science* (Cambridge: University Press, 1958), discusses retroduction. My findings suggest that the "H.-D. model" explains the facts of scientific discovery better than Hanson admits.

4. TAXONOMY

1 *Origin,* chap. 13; *Descent,* chap. 8.

2 *Orchids,* chap. 8. Cf. chap. 6 of this work.

[3] Although my taxonomic theory is eclectic, it has been greatly influenced by that of Mayr. For an introduction to taxonomy, I recommend E. Mayr, *Principles of Systematic Zoology* (New York: McGraw-Hill, 1969), and, with some reservations, G. G. Simpson, *Principles of Animal Taxonomy* (New York: Columbia University Press, 1961).

[4] *Origin*, p. 411.

[5] Perhaps the best known recent exposition of biological Platonism in English is A. Arber, *The Natural Philosophy of Plant Form* (Cambridge: University Press, 1950); for a zoological treatment, see A. Remane, *Die Grundlagen des natürlichen Systems, der vergleichenden Anatomie und der Phylogenetik*, 2d ed. (Leipzig: Geest & Portig, 1956).

[6] Cf. a letter of Darwin's dated 1861, in G. Sarton, "Darwin's Conception of the Theory of Natural Selection," *Isis*, XXVI (1937), 336–340.

[7] E.g., Mayr, "Introduction" to *Origin*.

[8] This topic is well treatedby D. L. Hull, "The Effect of Essentialism on Taxonomy—Two Thousand Years of Stasis (I)," *British Journal for the Philosophy of Science*, XV (1965), 314–326.

[9] *Origin*, p. 420.

[10] *Origin*, p. 486.

[11] J. S. Mill, *A System of Logic, Ratiocinative and Inductive, Being a Connected View of the Principles of Evidence and the Methods of Scientific Investigation*, 8th ed. (London: Longmans, 1874), pp. 466–467.

[12] D. L. Hull, "Consistency and Monophyly," *Systematic Zoology*, XIII (1964), 1–11.

[13] *Origin*, p. 420.

[14] M. P. Winsor, "Barnacle Larvae in the Nineteenth Century: A Case Study in Taxonomic Theory," *Journal of the History of Medicine and Allied Sciences* (in press).

[15] Attempts along this line have been made, for example, by R. F. Inger, "Comments on the Definition of Genera," *Evolution*, XII (1958), 370–384.

[16] *Second Notebook*, p. 91.

[17] This is where J. C. Buck and D. L. Hull, in "The Logical Structure of the Linnaean Hierarchy," *Systematic Zoology*, XV (1966), 97–111, 100, err in criticizing Mayr for speaking of certain taxa as "composed of" populations. Mayr referred, not to the taxonomic groups as classes, but to the real units in nature. The same mistake is made by J. R. Gregg in "Taxonomy, Language and Reality," *American Naturalist*, LXXXIV (1950), 419–425.

[18] In Wintu, for example; cf. H. Hoijer, "The Relation of Lan-

guage to Culture," *Anthropology Today: An Encyclopedic Inventory*, ed. A. L. Kroeber (Chicago: University Press, 1953), pp. 554–573, 571.

[19] A. Schopenhauer, *The Pessimist's Handbook: a Collection of Popular Essays* (Lincoln: University of Nebraska Press, 1964), p. 369.

[20] *More Letters*, I, 284.

[21] A similar argument is given by Simpson in *Animal Taxonomy* (above, n. 3), p. 69. Cf. also, M. T. Ghiselin, "An Application of the Theory of Definitions to Systematic Principles," *Systematic Zoology*, XV (1966), 127–130.

[22] E. Mayr, "Isolation as an Evolutionary Factor," *Proceedings of the American Philosophical Society*, CIII (1959), 221–230; G. de Beer, "Introduction" to *Second Notebook*.

[23] E.g., J. B. S. Haldane, "Can a Species Concept be Justified?" *The Species Concept in Paleontology*, ed. P. C. Sylvester-Bradley (London: Systematics Association, 1956), pp. 95–96.

[24] For an introduction to the literature, cf. works listed above, n. 3, and below, n. 37; also many articles in *Evolution* and *Systematic Zoology*.

[25] Use of an idealized model as a standard of definition has been stressed by J. Imbrie, in "The Species Problem in Fossil Animals," *The Species Problem*, ed. E. Mayr (Washington: American Association for the Advancement of Science, 1957), pp. 125–153. An essentialist interpretation is possible, but not unavoidable here.

[26] *Origin*, p. 485.

[27] Mayr, "Isolation."

[28] *Second Notebook*, p. 99. (The view expressed here was later modified.)

[29] *Life and Letters*, II, 333.

[30] *More Letters*, I, 453.

[31] C. Lyell, *Principles of Geology: Being an Inquiry How Far the Former Changes of the Earth's Surface are Referable to Causes Now in Operation*, 5th ed., 2 vols. (Philadelphia: James Kay, 1837), I, 481.

[32] Hull, "Effect of Essentialism" (above, n. 8).

[33] *Descent*, p. 175.

[34] *Origin*, p. 52.

[35] *Descent*, p. 176.

[36] *Origin*, p. 44.

[37] Cf. E. Mayr, "Species Concepts and Definitions," *The Species Problem*, ed. E. Mayr (Washington: American Association for the Advancement of Science, 1957), pp. 1–22.

[38] *Descent*, pp. 171–172.

[39] A number of modest proposals, based on the notion that such correspondence must exist, are made by T. M. Sonneborn, "Breeding Systems, Reproductive Methods, and Species Problems in Protozoa,"

The Species Problem, ed. E. Mayr (Washington: American Association for the Advancement of Science, 1957), pp. 155–324.

[40] *Origin*, p. 47.

[41] *Origin*, p. 52.

[42] *Origin*, p. 59.

[43] *First Notebook*, p. 44.

[44] *Variation*, II, 80; *Flowers*, p. 277.

[45] Cf. Mayr, "Isolation" (above, n. 22).

[46] *More Letters*, I, 212.

[47] *Origin*, chap. 13; the lexical accounts are in chap. 2.

[48] *Origin*, p. 411.

[49] *Origin*, p. 423.

[50] *Ibid.*

[51] *Origin*, p. 424.

[52] *Origin*, p. 485.

[53] Lyell, *Principles of Geology* (above, n. 31), I, 505–506 (includes earlier references).

[54] *Origin*, pp. 49–50.

[55] *Origin*, p. 485.

[56] *Flowers.*

[57] C. Darwin, "On the Specific Difference between *Primula veris*, Brit. Fl. (var. *officinalis* of Linn.), *P. vulgaris* Brit. Fl. (var. *acaulis*, Linn.), and *P. elatior*, Jacq., and on the Hybrid Nature of the Common Oxlip. With Supplementary Remarks on Naturally-produced Hybrids in the Genus *Verbascum*," *Journal of the Linnean Society* (*Botany*), X (1869), 437–454.

[58] *Ibid.*, p. 451.

[59] E. Mayr, "Difficulties and Importance of the Biological Species Concept," *The Species Problem*, ed. E. Mayr (Washington: American Association for the Advancement of Science, 1957), pp. 371–388.

[60] Simpson, *Animal Taxonomy* (above, n. 3), p. 153.

5. BARNACLES

[1] *Life and Letters*, I, 347–348.

[2] *Life and Letters*, I, 81–82.

[3] K. Goebel, "The Biology of Flowers," *Darwin and Modern Science*, ed. A. C. Seward (Cambridge: University Press, 1909), pp. 401–423; J. Heslop-Harrison, "Darwin as a Botanist," *A Century of Darwin*, ed. S. A. Barnett (Cambridge: Harvard University Press, 1959), pp. 267–295.

[4] N. Barlow, in "Darwin's Ornithological Notes," *Bulletin of the British Museum* (*Natural History*) *Historical Series*, II (1963), 201–278, 278, attributes this hypothesis to Sydney Smith.

[5] *First Notebook,* p. 56.

[6] C. Darwin, "Brief Descriptions of Several Terrestrial *Planariae,* and of some Remarkable Marine Species, with an Account of Their Habits," *Annals and Magazine of Natural History,* XIV (1844), 241–251. Darwin refers to a "Paper on new Balanus Arthrobalanus" in his journal; cf. G. de Beer, "Darwin's Journal," *Bulletin of the British Museum (Natural History) Historical Series,* II (1959), 1–200, 11; *Living Lepadidae,* p. v.

[7] *Life and Letters,* I, 346.

[8] *Life and Letters,* I, 80.

[9] *Life and Letters,* I, 81; also G. de Beer, "Some Unpublished Letters of Charles Darwin," *Notes and Records of the Royal Society of London,* XIV (1959), 12–66, 50.

[10] F. J. Cole, *A History of Comparative Anatomy* (London: Macmillan, 1944), fig. 3.

[11] W. Coleman, *Georges Cuvier, Zoologist: a Study in the History of Evolution Theory* (Cambridge: Harvard University Press, 1964).

[12] R. Magnus, *Goethe as a Scientist* (New York: Henry Schuman, 1949)—Platonic, inductionist, pro-Goethe; C. Sherrington, *Goethe on Nature and on Science* (Cambridge: University Press, 1949)—anti-Goethe and pro-Newton.

[13] *Living Balanidae,* p. 39.

[14] Biological Platonists generally.

[15] A. Boyden, "Homology and Analogy, a Critical Review of the Meanings and Implications of These Concepts in Biology," *American Midland Naturalist,* XXXVII (1947), 648–669.

[16] For examples, cf. M. T. Ghiselin, "An Application of the Theory of Definitions to Systematic Principles," *Systematic Zoology,* XV (1966).

[17] *Orchids,* chap. 8.

[18] M. P. Winsor, "Barnacle Larvae in the Nineteenth Century: A Case Study in Taxonomic Theory," *Journal of the History of Medicine and Allied Sciences* (in press).

[19] *Living Lepadidae,* p. 28.

[20] *Living Balanidae,* p. 151–152.

[21] *Living Lepadidae,* p. 38.

[22] P. Kruger, "Cirripedia," *H. G. Bronn's Klassen und Ordnungen des Tierreichs,* V (1:3:3) (1940), 106–112.

[23] *More Letters,* I, 64–65.

[24] J. Tomlinson, "The Advantages of Hermaphroditism and Parthenogenesis," *Journal of Theoretical Biology,* XI (1966), 54–58 treats this subject mathematically.

[25] A. R. Wallace, "On the Law Which has Regulated the Introduction of New Species," *Annals and Magazine of Natural History,* (2)

XVI (1855), 184–196, 196. For a good discussion on this and other aspects of Wallace's work, see C. F. A. Pantin, "Alfred Russel Wallace, F.R.S., and His Essays of 1858 and 1855," *Notes and Records of the Royal Society of London,* XIV (1959), 67–84.

26 *Fourth Notebook,* p. 165.

27 *Living Balanidae,* pp. 28–29.

28 *Third Notebook,* pp. 144–149; *Fourth Notebook,* p. 168.

29 *Descent,* pp. 161–164.

30 *Descent,* p. 161.

31 *Descent,* p. 162.

32 G. Bacci, *Sex Determination* (Oxford: Pergamon Press, 1965).

33 *Variorum,* pp. 682–683.

34 Cf. B. Patterson, "Rates of Evolution in Taeniodonts," *Genetics, Paleontology and Evolution,* ed. G. L. Jepsen, G. G. Simpson, and E. Mayr (Princeton: University Press, 1949), pp. 243–278.

35 The model for the difficulties of originating the hermaphroditic state is my own, and may or may not correspond to Darwin's. That for rudimentation may be documented as near to Darwin's own thinking.

36 M. T. Ghiselin, "Reproductive Function and the Phylogeny of Opisthobranch Gastropods," *Malacologia,* III (1966), 327–378.

37 For the beginnings of this train of thought, cf. *Third Notebook,* pp. 148–149.

38 G. de Beer, *Embryos and Ancestors,* 3d ed. (Oxford: Clarendon Press, 1958). For a critique of de Beer's views cf. R. A. Crowson, "Darwin and Classification," *A Century of Darwin,* ed. S. A. Barnett (Cambridge: Harvard University Press, 1959), pp. 102–129.

39 A. O. Lovejoy, "Recent Criticism of the Darwinian Theory of Recapitulation: Its Grounds and its Initiator," *Forerunners of Darwin,* ed. B. Glass, O. Temkin, and W. J. Strauss, Jr. (Baltimore: Johns Hopkins University Press, 1959), pp. 438–458.

40 A. Sedgwick, "The Influence of Darwin on the Study of Animal Embryology," *Darwin and Modern Science,* ed. A. C. Seward (Cambridge: University Press, 1909), pp. 171–184.

41 E. Haeckel, "Charles Darwin as an Anthropologist," *Darwin and Modern Science,* ed. A. C. Seward (Cambridge: University Press, 1909), pp. 137–151, 138.

42 E.g., E. Nordenskiöld, *The History of Biology* (New York: Tudor, 1928), chap. 14.

43 *Origin,* pp. 439–450.

44 G. de Beer, "Darwin's Views on the Relations Between Embryology and Evolution," *Journal of the Linnean Society of London (Zoology)* XLIV (1958), 15–23.

45 "Sketch," p. 78.

[46] Sedgwick, "Influence of Darwin on Animal Embryology," p. 174.

[47] T. Cahn, *La Vie et L'Oeuvre d'Étienne Geoffroy Saint-Hilaire* (Paris: Presses Universitaires de France, 1962), p. 237.

[48] *Fourth Notebook*, pp. 174–175; *Variation*, II, 379. See also de Beer, "Darwin's Views"; de Beer, *Charles Darwin: Evolution by Natural Selection* (London: Thomas Nelson & Sons, 1963), p. 218; and B. Klatt, "Darwin und die Haustierforschung," *Hundert Jahre Evolutionsforschung: das wissenschaftliche Vermächtnis Charles Darwins*, ed. G. Heberer and F. Schwanitz (Stuttgart: Gustav Fischer, 1960), pp. 149–168, 157.

[49] *Origin*, p. 441.

[50] *Living Balanidae*, pp. 15, 24, 26.

[51] *Living Lepadidae*, p. 180.

[52] *Living Lepadidae*, p. 200.

[53] *Living Lepadidae*, p. 221.

[54] *Living Balanidae*, pp. 439–440.

[55] *Living Lepadidae*, pp. 284–285.

[56] *Living Lepadidae*, p. 180.

[57] *Living Balanidae*, pp. 526–529.

[58] *Living Balanidae*, pp. 546–547.

[59] *Living Balanidae*, p. 535.

[60] *Living Lepadidae*, pp. 202–203.

[61] D. W. Thompson, *On Growth and Form* (Cambridge: University Press, 1917.

[62] Cf. chap. 7, "Variation Notes."

[63] G. R. de Beer, "Embryology and Taxonomy," *The New Systematics*, ed. J. Huxley (Oxford: University Press, 1940), pp. 365–393.

[64] E. Panofsky, *Meaning in the Visual Arts* (Garden City: Doubleday, 1955), chap. 2.

[65] *Life and Letters*, I, 70.

[66] C. Epling and W. Catlin, "The Relation of Taxonomic Method to an Explanation of Organic Evolution," *Heredity*, IV (1950), 313–325.

[67] M. T. Ghiselin, "William Harvey's Methodology in *De Motu Cordis*, from the Standpoint of Comparative Anatomy," *Bulletin of the History of Medicine*, XL (1966), 314–327.

6. A METAPHYSICAL SATIRE

[1] *First Notebook*, p. 41.

[2] *First Notebook*, p. 41.

[3] *Second Notebook*, p. 111.

[4] *Fourth Notebook*, p. 163.

[5] *Fourth Notebook*, p. 179.

[6] For the same reason that he misses the significance of Malthus, G. de Beer, in "Introduction" to *Fourth Notebook*, p. 155, unfairly criticizes Darwin's views on the so-called final cause of sex.

[7] *Naturalist's Voyage* (1840 ed.), p. 262.

[8] R. C. Olby, *Origins of Mendelism* (London: Constable, 1966).

[9] *Life and Letters*, I, 385.

[10] *Life and Letters*, I, 81.

[11] *Living Lepadidae*, p. 203.

[12] J. Dewey, *The Influence of Darwin on Philosophy and Other Essays in Contemporary Thought* (New York: Henry Holt and Company, 1910).

[13] M. Beckner, *The Biological Way of Thought* (New York: Columbia University Press, 1959); E. Nagel, "Teleological Explanation and Teleological Systems," *Readings in the Philosophy of Science*, ed. H. Feigl and M. Brodbeck (New York: Appleton-Century-Crofts, 1953), pp. 537–558.

[14] *Life and Letters*, I, 47.

[15] W. Paley, *Natural Theology: or, Evidences of the Existence and Attributes of the Deity, Collected from the Appearances of Nature* (Boston: Gould and Lincoln, 1851).

[16] A. Gray, "Design Versus Necessity, Discussion Between Two Readers of Darwin's Treatise on the Origin of Species, Upon its Natural Theology," A. Gray, *Darwiniana, Essays and Reviews Pertaining to Darwinism* (Cambridge: Harvard University Press, 1963) (original date of article: 1860), pp. 51–71. For Gray's relationship with Darwin, I have found A. H. Dupree's introduction to this work and his *Asa Gray:1810–1888* (Cambridge: Harvard University Press, 1959) exceedingly useful.

[17] Dewey, *Influence of Darwin on Philosophy*.

[18] G. Himmelfarb, in *Darwin and the Darwinian Revolution* (New York: Doubleday, 1959), p. 330, blunders in a manner best compared to calling Voltaire a Pangloss.

[19] *Life and Letters*, II, 266.

[20] *More Letters*, I, 202.

[21] Aristotle, *Physics, De Partibus Animalium*.

[22] W. J. Bock and G. von Wahlert, in "Adaptation and the Form-function Complex," *Evolution*, XIX (1965), 269–299, 274, try to get around the difficulties by redefining "function" to mean an intrinsic property, namely, a physical or chemical property arising from form.

[23] For a more extensive treatment, cf. H. Lehman, "Functional Explanation in Biology," *Philosophy of Science*, XXXII (1965), 1–20.

[24] Paley, *Natural Theology* (above, n. 10), p. 10.

[25] I cannot agree with Lehman ("Functional Explanation," p. 16) that function statements presuppose the notion of a cause.

[26] Cf. chap. 1, "Geology."

[27] Cf. chap. 7, "Variation."

[28] Gray, *Darwiniana* (above, n. 16), pp. 69–70.

[29] *Orchids,* chaps. 2, 3.

[30] *Orchids,* pp. 162–166.

[31] *Orchids,* pp. 36–44.

[32] *Flowers,* p. 15.

[33] *Life and Letters,* I, 91.

[34] A most useful review is H. L. K. Whitehouse, "Cross- and Self-fertilization in Plants," *Darwin's Biological Work, Some Aspects Reconsidered,* ed. P. R. Bell (Cambridge: University Press, 1959), pp. 207–261.

[35] *Life and Letters,* I, 91.

[36] *Life and Letters,* I, 91; *Flowers,* p. 18.

[37] *Flowers,* p. 19.

[38] *Flowers,* pp. 22–24.

[39] *Flowers,* pp. 24–30.

[40] A failure to make this distinction has led to misinterpretations of Harvey's functional anatomy; cf. M. T. Ghiselin, "William Harvey's Methodology in *De Motu Cordis,* from the Standpoint of Comparative Anatomy," *Bulletin of the History of Medicine,* XL (1966), 314–327.

[41] E. Mayr, *Animal Species and Evolution* (Cambridge: Harvard University Press, 1963).

[42] T. Dobzhansky, *Genetics and the Origin of Species,* 3d ed. (New York: Columbia University Press, 1951), pp. 204–206.

[43] E. Mayr, "Isolation as an Evolutionary Factor," *Proceedings of the American Philosophical Society,* CIII (1959), 221–230.

[44] A. O. Lovejoy, "Buffon and the Problem of Species," *Forerunners of Darwin: 1745–1859,* ed. B. Glass, O. Temkin, and W. L. Strauss (Baltimore: Johns Hopkins Press, 1959) pp. 84–113, 101.

[45] Mayr, *Animal Species and Evolution.*

[46] *Fertilisation,* p. 9.

[47] *Fertilisation,* p. 382.

[48] This matter is further discussed in *Variation,* chap. 20.

[49] D. Amadon, "The Evolution of Low Reproductive Rates in Birds," *Evolution,* XVIII (1964), 105–110.

[50] W. D. Hamilton, "The Genetical Evolution of Social Behaviour, I, II," *Journal of Theoretical Biology,* VII (1964), 1–16, 17–52.

[51] *More Letters,* I, 288–299; cf. *Variorum,* pp. 443–459.

[52] *More Letters,* I, 297.

[53] *More Letters*, I, 296.

[54] *Flowers*, p. 139.

[55] *Flowers*, p. 161.

[56] *Orchids*, pp. 257–262.

[57] *Orchids*, pp. 202–203; the botanist was de Candolle—cf. "Essay," p. 240.

[58] *Orchids*, pp. 283–284.

[59] *Orchids*, p. 284.

[60] *Descent*, p. 441.

[61] A. Gray, "Natural Selection Not Inconsistent With Natural Theology," Gray, *Darwiniana* (above, n. 16) (original date of article: 1860), pp. 72–145, 121–122.

[62] *Variation* II, 426–448.

[63] *Variation* II, 427–428.

[64] B. L. Whorf, "Languages and Logic," *Technological Review*, XLIII (1941), 250–252, 266, 268, 272.

7. VARIATION

[1] Mayr, "Introduction" to *Origin;* A. Müntzing, "Darwin's Views on Variation under Domestication in the Light of Present-day Knowledge," *Proceedings of the American Philosophical Society*, CIII (1959), 190–220.

[2] R. A. Fisher, "Has Mendel's Work Been Rediscovered?" G. Mendel, *Experiments in Plant Hyridisation, Mendel's Original Paper in English Translation, with Commentary and Assessment by the Late Sir Ronald A. Fisher, M. A., Sc. D., F. R. S., together with a Reprint of W. Bateson's Biographical Notice of Mendel*, ed. J. H. Bennett (Edinburgh and London: Oliver & Boyd, 1965), pp. 59–87; R. C. Olby, *Origins of Mendelism* (London: Constable, 1966).

[3] Cf. E. B. Gasking, "Why Was Mendel's Work Ignored?" *Journal of the History of Ideas*, XX (1959), 60–84.

[4] R. A. Fisher, *The Genetical Theory of Natural Selection* (Oxford: Clarendon Press, 1930), p. 36.

[5] E.g., C. D. Darlington, *Darwin's Place in History* (Oxford: Blackwell, 1960); and to some extent L. Eisley, *Darwin's Century: Evolution and the Men Who Discovered It* (New York: Anchor, 1961), pp. 244–246, etc.

[6] E.g., G. de Beer, *Charles Darwin: Evolution by Natural Selection* (London: Thomas Nelson & Sons, 1963), pp. 89–90.

[7] Darwin does invoke this argument (*Origin*, p. 105) but finds it impossible to verify.

[8] *Descent*, p. 181.

[9] *Variorum*, pp. 121, 264–265.

[10] *Variation,* I, 4.

[11] *Variation,* I, 102.

[12] *Descent,* p. 469.

[13] *Variation,* II, 282.

[14] *Fourth Notebook,* 164.

[15] De Beer, footnote in *Fourth Notebook,* p. 164.

[16] *Descent,* p. 230.

[17] *Variation,* II, 69–72.

[18] *Origin,* pp. 111–126; *Variation,* II, 157; *Descent,* p. 172.

[19] *Insectivorous Plants.*

[20] *Insectivorous Plants,* p. 173.

[21] G. G. Simpson, "Anatomy and Morphology: Classification and Evolution: 1859 and 1959," *Proceedings of the American Philosophical Society,* CIII (1959), 286–306, 297.

[22] *Variation,* II, 51.

[23] *Variation,* I, 121–126.

[24] *Variation,* I, 155.

[25] The fourth notebook on transmutation of species gives Cuvier as a source for this idea; cf. G. de Beer and M. J. Rowlands, "Darwin's Notebooks on Transmutation of Species. Addenda and Corregenda," *Bulletin of the British Museum (Natural History), Historical Series,* II (1961), 185–200, 195.

[26] *Variation,* I, 36.

[27] *Variation,* I, 105; II, 318.

[28] *Variation,* II, 186.

[29] *Variation,* II, 319–322.

[30] Simpson, "Anatomy and Morphology" (above, n. 21), 293; E. Mayr, in *Animal Species and Evolution* (Cambridge: Harvard University Press), pp. 2, 176, 281–282, stresses the fact that the more recent "Modern Synthesis" has come to recognize the importance of "mutational limitations" and "epigenetic limitations," and these are clearly the sort of phenomenon that Darwin had in mind.

[31] *Variation,* I, 58.

[32] *Variation,* I, 58.

[33] *Descent,* p. 385.

[34] Mayr, *Animal Species and Evolution,* p. 281.

[35] *Origin,* pp. 146–147.

[36] *Variation,* II, 322–323.

[37] *Variation,* II, 32.

[38] E.g., *Variation,* I, 229–230.

[39] P. Vorzimmer, "Charles Darwin and Blending Inheritance," *Isis,* LIV (1963), 371–390.

[40] *Ibid.,* p. 390.

[41] *Variation,* II, 157.

⁴² *Variation*, II, 245.

⁴³ *Fertilisation*, pp. 8–9.

⁴⁴ *Fertilisation*, pp. 15–19, 146, 234–235.

⁴⁵ *Fertilisation*, p. 266.

⁴⁶ *Fertilisation*, p. 236.

⁴⁷ Cf. F. Brabec, "Darwin's Genetik im lichte der modernen Verer-
bungslehre erläutert an botanischen Beispielen," G. Heberer and F.
Schwanitz, *Hundert Jahre Evolutionsforschung: das wissenschaftliche
Vermächtnis Charles Darwins* (Stuttgart: Gustav Fischer Verlag, 1960),
pp. 99–122; H. L. K. Whitehouse, "Cross- and Self-fertilization in
plants," *Darwin's Biological Work: Some Aspects Reconsidered*, ed.
P. R. Bell (Cambridge: University Press, 1959), pp. 207–261.

⁴⁸ *Fertilisation*, p. 254.

⁴⁹ *Variation*, II, 161.

⁵⁰ *Variation*, I, 311.

⁵¹ *Third Notebook*, p. 145; *Descent*, pp. 227–228; *Variation* II, 25–
27.

⁵² *Origin*, pp. 159–167; *Variation*, II, 25–36.

⁵³ *Variation*, II, 10–12.

⁵⁴ *Variation*, II, 25.

⁵⁵ H. Steiner, "Atavismen bei Artbastarden und ihre Bedeutung zur
Feststellung von Verwandtschaftsbeziehungen. Kreuzungsergenbnisse
innerhalb der Singvogelfamilie der *Spermestidae*," *Revue Suisse de
Zoologie*, LXXIII (1966), 321–337.

⁵⁶ *Variation*, II, 22–23.

⁵⁷ The "act of crossing" refers not to the mere act, but to its effect
on morphogenetic processes; cf. Variation, chap. 7.

⁵⁸ *Variation*, II, 45.

⁵⁹ *Descent*, p. 181.

⁶⁰ "Sketch"; "Essay."

⁶¹ *Descent*, pp. 565–566.

⁶² *Variation*, I, 467–469; *Variorum*, p. 281.

⁶³ *Variation*, I, 469.

⁶⁴ *Variation*, I, 470.

⁶⁵ *Variation*, II, 289.

⁶⁶ *Variation*, II, 309–310.

⁶⁷ *Variorum*, p. 75.

⁶⁸ C. Zirkle, "The Early History of the Inheritance of Acquired
Characters and of Pangenesis," *Transactions of the American Philosophical Society*, XXXV (1946), 91–151, 119.

⁶⁹ As has been pointed out by C. Stern in "Variation and Hereditary
Transmission," *Proceedings of the American Philosophical Society*,
CIII (1959), 183–189, and Mayr, in "Introduction" to *Origin*, p. xxvi.

⁷⁰ *Variation*, II, 370.

[71] *Naturalist's Voyage* (1845 ed.), p. 52 (not in earlier versions); for Darwin's changing views on use and disuse, see *Variorum,* pp. 713–716.

[72] *Variation,* II, 307–309; C. Darwin, "On the Males and Complemental Males of Certain Cirripedes, and on Rudimentary Structures," *Nature,* VIII (1873), 431–432; C. Darwin, "Fritz Müller on a Frog Having Eggs on its Back—on the Abortion of the Hairs on the Legs of Certain Caddis-flies, etc.," *Nature,* XIX (1879), 462–463.

[73] Fisher, *Genetical Theory* (above, n. 4), p. 20.

[74] *Variation,* II, 388–389, 392–393.

[75] J. Heslop-Harrison, "Darwin as a Botanist," *A Century of Darwin,* ed. S. A. Barnett (Cambridge: Harvard University Press, 1959), p. 289; Müntzing, "Darwin's Views on Variation" (above, n. 1).

[76] As in the title of C. D. Darlington, "Purpose and Particles in the Study of Heredity," *Science, Medicine and History . . . ,* ed. E. Ashworth Underwood (Oxford: University Press, 1953); cf. chap. 6, "A Metaphysical Satire."

[77] Darlington, *Darwin's Place* (above, n. 5), p. 35.

[78] *Ibid.,* p. 40.

8. AN EVOLUTIONARY PSYCHOLOGY

[1] A recent example: L. S. Hearnshaw, *A Short History of British Psychology 1840–1900* (London: Methuen, 1964).

[2] G. de Beer, *Charles Darwin: Evolution by Natural Selection* (London: Thomas Nelson & Sons, 1963), pp. 249–250, gives this impression; also W. Irvine, *Apes, Angels, and Victorians* (New York: McGraw Hill, 1955), p. 203.

[3] M. Mead, "Preface" to *Expression.*

[4] P. Marler, "Developments in the Study of Animal Communication," *Darwin's Biological Work: Some Aspects Reconsidered,* ed. P. R. Bell (Cambridge: University Press, 1959), pp. 150–206.

[5] As has been pointed out by I. Eibl-Eibesfeldt in "Darwin und die Ethologie," *Hundert Jahre Evolutionsforschung, das wissenschaftliche Vermächtnis Charles Darwins,* ed. G. Heberer and F. Schwanitz (Stuttgart: Gustav Fischer, 1960), pp. 355–367, 365.

[6] *Descent,* p. 402.

[7] Contrary to the view of S. A. Barnett, "The 'Expression of the Emotions,'" *A Century of Darwin,* ed. S. A. Barnett (Cambridge: Harvard University Press, 1959), pp. 206–230; cf. Hearnshaw, *Short History of British Psychology,* p. 39.

[8] *Climbing Plants,* p. 99.

[9] *Climbing Plants,* p. 6.

[10] *Expression,* figs. 9 and 10.

¹¹ Barnett, "The 'Expression of the Emotions,' " p. 225.

¹² J. B. Watson, *Behaviorism* 2d ed. (Chicago: University Press, 1930), p. 238.

¹³ *Ibid.*, pp. 103–105.

¹⁴ J. Hirsch, "Behavior Genetics and Individuality Understood," *Science*, CXLII (1963), 1436–1442, 1436.

¹⁵ *Variation*, I, 156–161; C. Darwin, "Origin of Certain Instincts," *Nature*, VII (1873), 417–418.

¹⁶ C. Zirkle, *Evolution, Marxian Biology and the Social Scene* (Philadelphia: University of Pennsylvania Press, 1959).

¹⁷ Cf. H. Feigl, "Logical Empiricism," *Readings in Philosophical Analysis*, ed. H. Feigl and W. Sellars (New York: Appleton-Century-Crofts, 1949), pp. 7–26, 10.

¹⁸ J. H. Woodger, *Biology and Language: an Introduction to the Methodology of the Biological Sciences, Including Medicine* (Cambridge: University Press, 1952).

¹⁹ S. Toulmin, "Are the Principles of Logical Empiricism Relevant to the Actual Work of Science?" (review article), *Scientific American*, CCXIV (1966), 129–133.

²⁰ C. Lloyd Morgan, *An Introduction to Comparative Psychology* (London: Walter Scott, 1894), p. 53.

²¹ K. W. Spence, "The Postulates and Methods of 'Behaviorism,' " *Readings in the Philosophy of Science*, ed. H. Feigl and M. Brodbeck (New York: Appleton-Century-Crofts, 1953), pp. 571–584.

²² *Climbing Plants.*

²³ *Naturalist's Voyage* (1845 ed.), pp. 44–45, 111.

²⁴ *Climbing Plants*, pp. 189–206.

²⁵ Cf. F. A. Beach, "The Snark was a Boojum," *The American Psychologist*, V (1950), 115–124. Darwin's importance is often overlooked by workers in this field.

²⁶ *Origin*, pp. 219–235.

²⁷ P. R. Bell, "The Movement of Plants in Response to Light," *Darwin's Biological Work, Some Aspects Reconsidered*, ed. P. R. Bell (Cambridge: University Press, 1959), pp. 1–49, 6–7.

²⁸ F. Darwin, "Darwin's Work on the Movements of Plants," *Darwin and Modern Science*, ed. A. C. Seward (Cambridge: University Press, 1909), pp. 385–400.

²⁹ *Ibid.*, p. 389.

³⁰ L. M. Passano, "Primitive Nervous Systems," *Proceedings of the National Academy of Sciences*, L (1963), 306–313, invokes a model, strikingly similar to Darwin's, to explain certain features of evolution in animal nervous systems.

³¹ Cf. above, n. 2.

[32] F. Darwin, "Darwin's Work on the Movements of Plants," pp. 391–392; *Climbing Plants*, pp. 179–182.

[33] *Movement*, pp. 112–122.

[34] *Movement*, p. 122.

[35] *Movement*, p. 123.

[36] *Movement*, pp. 484–489, Bell, "Movement of Plants" (above, n. 27), pp. 5–7.

[37] *Movement*, chap. 9.

[38] Cf. E. Hauenstein, "Darwin als Botaniker," *Hundert Jahre Evolutionsforschung: das wissenschliche Vermächtnis Charles Darwins*, ed. G. Heberer and F. Schwanitz (Stuttgart: Gustav Fischer, 1960), pp. 168–185.

[39] *Life and Letters*, I, 95.

[40] *Insectivorous Plants*, chap. 6.

[41] *Insectivorous Plants*, chap. 9.

[42] *Insectivorous Plants*, pp. 186–188.

[43] *Insectivorous Plants*, p. 3.

[44] *Insectivorous Plants*, pp. 247–253.

[45] *Insectivorous Plants*, p. 240.

[46] *Life and Letters*, III, 322.

[47] *Movement*, chap. 3; cf. Hauenstein, "Darwin als Botaniker" (above, n. 38).

[48] *Insectivorous Plants*, chaps. 2, 3.

[49] *Movement*, pp. 559–573.

[50] *Movement*, p. 573.

[51] C. Darwin, "On the Formation of Mould," *Proceedings of the Geological Society of London*, II (1838), 574–576.

[52] *Life and Letters*, III, 243.

[53] *Worms*, p. 24.

[54] *Worms*, p. iv.

[55] *Worms*, p. 74.

[56] *Worms*, pp. 75–77.

[57] *Worms*, pp. 88–90.

[58] *Worms*, p. 90.

[59] E. Hanel, "Ein Beitrag zur 'Psychologie' der Regenwürmer," *Zeitschrift für allgemeine Physiologie*, IV (1904), 244–258.

[60] H. Jordan, "Wie ziehen die Regenwürmer Blätter in ihre Röhren? Ein Beitrag zur Physiologie der nahrungsaufnehmenden Organe und zur Psychologie der Regenwürmer," *Zoologische Jahrbücher, Abteilung für allgemeine Zoologie und Physiologie der Tiere*, XXXIII (1912), 95–106; O. Mangold, "Beobachtungen und Experimente zur Biologie des Regenwurms. I," *Zeitschrift für vergleichende Physiologie*, II (1924), 57–81.

[61] R. M. Yerkes, "The Intelligence of Earthworms," *Journal of Animal Behavior*, II (1912), 322–352; J. O. Krivanek, "Habit Formation in the Earthworm *Lumbricus terrestris*," *Physiological Zoology*, XXIX (1956), 241–250.

[62] E.g., H. Schmidt, "Behavior of Two Species of Worms in the Same Maze," *Science*, CXXI (1955), 341–342.

[63] Above, notes 3, 4.

[64] Marler, "Developments in the Study of Animal Communication" (above, n. 4).

[65] *Expression*, chaps. 1, 2, and 3, respectively.

[66] T. H. Waterman, "Systems Analysis and the Visual Orientation of Animals," *American Scientist*, LIV (1966), 15–45, 16–17.

[67] *Expression*, pp. 220, 336; *Life and Letters*, III, 96; *Descent*, pp. 3–4.

[68] Marler, in "Developments . . . Animal Communication," pp. 199–200, recognizes the nonadaptive nature of movements resulting from direct action and gives further references to the literature.

[69] *Expression*, p. 169.

[70] *Expression*, pp. 49, 126.

[71] C. Lloyd Morgan, "Mental Factors in Evolution," *Darwin and Modern Science*, ed. A. C. Seward (Cambridge: University Press, 1909), pp. 424–445, 433.

[72] Marler, "Developments . . . Animal Communication," p. 198.

[73] *Expression*, p. 344.

[74] *Expression*, p. 349.

[75] *Expression*, p. 118.

[76] *Expression*, p. 127.

[77] C. Darwin, "A Posthumous Essay on Instinct," appendix to G. J. Romanes, *Mental Evolution in Animals* (New York: D. Appleton, 1883), pp. 355–384, 376.

[78] *Expression*, p. 122.

[79] *Expression*, pp. 6–7.

[80] *Descent*, p. 402.

[81] *Descent*, p. 67.

[82] *Descent*, pp. 67–68.

[83] *Origin*, pp. 219–242.

[84] *Origin*, p. 242.

[85] *Origin*, p. 209.

[86] E.g., F. A. Beach, "The Descent of Instinct," *Psychological Review*, LXII (1955), 401–410; W. H. Thorpe, *Learning and Instinct in Animals*, 2d ed. (London: Methuen, 1963).

[87] For a very readable treatment of the problem see T. Dobzhansky, *Mankind Evolving: the Evolution of the Human Species* (New Haven: Yale University Press, 1962), chap. 3.

[88] *Origin*, pp. 207–208.

[89] J. Dewey, "The Theory of Emotion. I," *Psychological Review*, I (1894), pp. 553–569, 562, 566–567.

[90] H. W. Magoun, "Evolutionary Concepts of Brain Function Following Darwin and Spencer," *Evolution After Darwin*, ed. S. Tax (Chicago: University Press, 1960) II, 187–218.

[91] R. J. Herrnstein, "Superstition: A Corollary of the Principles of Operant Conditioning," *Operant Behavior, Areas of Research and Application*, ed. W. K. Honig (New York: Appleton-Century-Crofts, 1966), pp. 33–51; B. F. Skinner, "The Phylogeny and Ontogeny of Behavior," *Science*, CLIII (1966), pp. 1205–1213.

[92] Cf. Thorpe, *Learning and Instinct*.

[93] K. Lorenz, "The Comparative Method in Studying Innate Behavior Patterns," *Symposia of the Society for Experimental Biology*, IV (1950), 221–268, 232. In his "Gestaltwahrnehmung als Quelle wissenschaftlicher Erkenntnis," *Zeitschrift für experimentelle und Angewandte Psychologie*, VI (1959), 118–165, Lorenz, confusing the logically and psychologically prior, invokes evolution to support the Platonism which pervades his writings.

[94] M. Mead, "Preface" to *Expression*.

9. SEXUAL SELECTION

[1] *Origin*, p. 88.

[2] *Origin*, pp. 89–90.

[3] G. Schwalbe, " 'The Descent of Man,' " *Darwin and Modern Science*, ed. A. C. Seward (Cambridge: University Press, 1909), pp. 112–136, 125. L. von Bertallanfy, in *Problems of Life, an Evaluation of Modern Biological Thought* (New York: John Wiley & Sons, 1962), p. 89, charges this of selection theory in general. G. Himmelfarb, in *Darwin and the Darwinian Revolution* (New York, Doubleday, 1959), pp. 298–302, 317, 319, 343–346, uses this notion as one of her basic premises.

[4] G. de Beer, *Charles Darwin: Evolution by Natural Selection* (London: Thomas Nelson & Sons, 1963), pp. 219–221.

[5] P. O'Donald, "The Theory of Sexual Selection," *Heredity*, XVII (1962), 541–552.

[6] J. S. Huxley, "The Present Standing of the Theory of Sexual Selection," *Evolution: Essays on Aspects of Evolutionary Biology Presented to Professor E. S. Goodrich on his Seventieth Birthday*, ed. G. R. de Beer (Oxford: Clarendon Press, 1938), pp. 11–42 (not reliable for Darwin's views); J. M. Smith, "Sexual Selection," *A Century of Darwin*, ed. S. A. Barnett (Cambridge: Harvard University Press, 1959), pp. 231–244.

[7] E.g., Nordenskiöld, *History of Biology* (New York: Tudor, 1928), p. 474.

[8] *Descent*, pp. 207–210.

[9] *Descent*, p. 209.

[10] *Descent*, p. 210.

[11] *Descent*, pp. 509–510.

[12] Nordenskiöld, *History of Biology*, p. 474.

[13] *Descent*, p. 211. Cf. C. Darwin and W. Van Dyck, "On the Modification of a Race of Syrian Street-dogs by Means of Sexual Selection," *Proceedings of the Zoological Society of London*, no vol. no. (1882), pp. 367–370.

[14] G. de Beer, *Charles Darwin* (above, n. 4), p. 219; L. Eiseley, *Darwin's Century* . . . (New York: Anchor, 1961), p. 47; C. D. Darlington, *Darwin's Place in History* (Oxford: Blackwell, 1960), p. 11; W. Irvine, *Apes, Angels, and Victorians* (New York: McGraw-Hill, 1955), p. 85.

[15] N. Barlow, *Autobiography of Charles Darwin 1809–1882* (New York: Harcourt, Brace, 1959), pp. 149–166; E. Darwin, *The Temple of Nature; or, the Origin of Society: a Poem, with Philosophical Notes* (New York: Swords, 1804), pp. 172–173, 179; an entire series of misconceptions about the relationships between the two Darwins are given by D. King-Hele, *Erasmus Darwin* (New York: Charles Scribner's Sons, 1964).

[16] *Descent*, p. 211.

[17] *Third Notebook*, pp. 139, 145, 148.

[18] "Sketch," pp. 48–49.

[19] "Essay," pp. 120–121.

[20] C. Darwin and A. R. Wallace, *Evolution by Natural Selection* (Cambridge: University Press, 1958), p. 263.

[21] *Origin*, pp. 87–90.

[22] *Life and Letters*, III, 90–97.

[23] *Descent*, chap. 9.

[24] *Descent*, p. 264.

[25] *Descent*, p. 263.

[26] The differences are primarily behavioral. Cf. M. J. Wells, *Brain and Behaviour in Cephalopods* (Stanford: University Press, 1962), pp. 33–37.

[27] *Descent*, p. 261.

[28] *Variation*, I, 139; II, 202–205.

[29] *Ibid.*

[30] *Descent*, pp. 260–555.

[31] *Descent*, p. 314.

[32] *Descent*, pp. 434–442.

[33] *Descent*, p. 441.

[34] *Descent*, p. 326.

[35] *Descent*, pp. 324–325.

[36] P. M. Sheppard, "The Evolution of Mimicry: A Problem in Ecology and Genetics," *Cold Spring Harbor Symposia on Quantitative Biology*, XXIV (1959), 131–140.

[37] *Descent*, p. 324; cf. R. B. Goldschmidt, "Mimetic Polymorphism, a Controversial Chapter of Darwinism," *Quarterly Review of Biology*, XX (1945), 147–164, 205–230.

[38] Sheppard, "Evolution of Mimicry."

[39] *Descent*, p. 324.

[40] *Descent*, p. 325.

[41] E. B. Ford, "The Genetics of Polymorphism in the Lepidoptera," *Advances in Genetics*, V (1953), 43–87, 68.

[42] Cf. W. George, *Biologist Philosopher, A Study of the Life and Writings of Alfred Russel Wallace* (London: Abelard-Schuman, 1964), pp. 183–198.

[43] C. J. Duncan and P. M. Sheppard, "Sensory Discrimination and its Role in the Evolution of Batesian Mimicry," *Behaviour*, XXIV (1965), 269–282; R. S. Schmidt, "Predator Behaviour and the Perfection of Incipient Mimetic Resemblances," *Behaviour*, XVI (1960), 149–158.

[44] *Descent*, pp. 216–221.

[45] Himmelfarb, *Darwin* (above, n. 3).

[46] *Descent*, pp. 213–215.

[47] Cf. *Descent*, p. 591.

[48] *Descent*, chaps. 15, 16.

[49] *Variorum*, p. 372. (This argument was omitted in the sixth edition).

[50] *More Letters*, II, 59–93.

[51] *Descent*, p. 444.

[52] *Descent*, p. 506.

[53] *Descent*, pp. 461–462.

[54] *Descent*, pp. 449–450.

[55] *Descent*, p. 452.

[56] *Third Notebook*, p. 139.

[57] Cf. chap. 8, "An Evolutionary Psychology."

[58] *Descent*, p. 466.

[59] J. Huxley, "The Emergence of Darwinism," *Journal of the Linnaean Society of London*, LVI and XLIV (1958), 1–14.

[60] C. Darwin, "Sexual Selection in Relation to Monkeys," *Nature*, XV (1876), 18–19.

[61] *Life and Letters*, III, 89–95; *More Letters*, II, 56–97.

[62] For an example of one who completely missed the point, cf. E. B. Poulton, "The Value of Colour in the Struggle for Life," *Darwin and*

Modern Science, ed. A. C. Seward (Cambridge: University Press, 1909), pp. 271–297.

[63] Descent, p. 597.

[64] G. K. Noble, "Sexual Selection among Fishes," Biological Reviews, XIII (1938), 133–158; L. Ehrman, B. Spassky, O. Pavlovsky, and T. Dobzhansky, "Sexual Selection, Geotaxis, and Chromosomal Polymorphism in Experimental Populations of Drosophila Pseudoobscura," Evolution, XIX (1965), 337–346.

[65] E. Mayr, Animal Species and Evolution (Cambridge: Harvard University Press, 1963), pp. 126–127; N. Tinbergen, "The Origin and Evolution of Courtship and Threat Display," Evolution as a Process, ed. J. Huxley, A. C. Hardy, and E. B. Ford (London: George Allen and Unwin, 1954), pp. 233–250.

[66] G. K. Nobel and H. T. Bradley, "The Mating Behavior of Lizards: Its Bearing on the Theory of Sexual Selection," Annals of the New York Academy of Sciences, XXXV (1933), 25–100.

[67] E. Eisner, "The Relationship of Hormones to the Reproductive Behaviour of Birds, Referring Especially to Parental Behaviour: A Review," Animal Behaviour, VIII (1960), 155–179.

10. CONCLUSIONS

[1] C. C. Gillispie, "Lamarck and Darwin in the History of Science," American Scientist, XLVI (1958), 388–409, 389. The whole statement is even more curious: "More generally, it might be argued—indeed, I do argue—that in the relative cogency with which the two theories organize actual biological information, Lamarck's presentation in the great Histoire naturelle des animaux sans vertèbres is the more interesting and elegant. It is analytical and informs a systematic taxonomy, whereas Darwin simply amassed detail and pursued his argument through the accumulated observations in a naturalist's commonplacebook."

[2] Cf. N. R. Hanson, Patterns of Discovery . . . (Cambridge: University Press, 1958); H. A. Simon and A. Newell, "Information Processing in Computer and Man," American Scientist, LII (1964), 281–300.

[3] Third Notebook, p. 137.

[4] Life and Letters, I, 107; III, 179.

[5] Life and Letters, I, 102, 103.

[6] Life and Letters, I, 33, 47.

[7] D. T. Campbell, "Blind Variation and Selective Retention in Creative Thought as in Other Knowledge Processes," Psychological Review, LXVII (1960), 380–400.

[8] Life and Letters, I, 69.

9 P. H. Barrett, "From Darwin's Unpublished Notebooks," *The Centennial Review*, III (1959), 391–406, 403.

10 L. Eiseley, "Darwin, Coleridge, and the Theory of Unconscious Creation," *Daedalus*, XCIII (1965), 588–602, 597; I note that a quotation (p. 596) deletes some crucial words from Darwin's statement about his fellow naturalists' views—cf. *Life and Letters*, I, 87.

11 Cited in Eiseley, "Darwin, Coleridge."

12 Aristotle, *De Partibus Animalium*; R. Magnus, *Goethe as a Scientist* (New York: Henry Schuman, 1949); *Variation*, II, 335.

13 Aristotle, *Parts of Animals*, A. L. Peck, trans. (Cambridge: Harvard University Press, 1955), p. 183.

14 *Variation*, II, 335–336.

15 L. Eiseley, *Darwin's Century* . . . (New York: Anchor, 1961), pp. 341–342. On the basis of evidence given in chap. 6, I come to conclusions diammetrically opposite to Eiseley's view (p. 342) that "it is difficult to find in Darwin any really deep recognition of the life of the organism as a functioning whole which must be co-ordinated interiorly before it can function exteriorly."

16 F. Bacon, "The Advancement of Learning," *Selected Writings of Francis Bacon*, ed. H. G. Dick (New York: Modern Library, 1955), p. 182.

17 J. C. Greene, *The Death of Adam: Evolution and its Impact on Western Thought* (New York: Mentor, 1959), pp. 255, 257, 366.

18 J. C. Greene, *Darwin and the Modern World View* (New York· Mentor, 1963), p. 70.

19 *Life and Letters*, I, 107.

20 Some references are given in *Life and Letters*.

21 *Life and Letters*, I, 66–67.

Index

Acquired characteristics, 140
Ad hoc hypothesis: in scientific method, 17, 20; erroneous charges of, 31, 39-40, 162, 180, 184, 215; by Darwin's critics, 216, 239; mentioned, 173
Adaptation: semantics of, 55, 64-65, 73; biology of, 56-57, 133-134, 211; and progress, 70-71; and teleology, 156, 205, 215; mentioned, 75, 132, 136, 229
Agassiz, Alexander, 26
Agassiz, Louis, 42-43, 45, 81, 92, 121, 136
Analogy: in discovery *vs.* verification, 59, 146, 239; between sciences, 67, 68, 74, 233; morphological, 109, 116; Darwin's use, 118, 127, 145, 156, 168, 200-201, 207; mentioned, 121, 134, 163, 199, 208, 219
Anthropology, 187-188, 212-213, 242-243
Anthropomorphism, 71, 188-189, 201, 203, 218, 241
Antithesis, 204, 206, 207
Archetype. *See* Essence
Aristotelianism: and definition, 51, 52-53; and classification, 70, 84, 89, 95; in Darwin's thought, 81-82, 92; mentioned, 55, 240
Aristotle: on definition, 50-52; on potentiality, 55; on form, 106, 107, 171, 239-240; on teleology, 137, 138
Artificial selection: argument for evolution, 4, 160; in discovery of natural selection, 48-49; in Darwin's evolutionary theory, 163, 165, 172; in discovery of sexual selection, 218-220, 221; mentioned, 183, 230, 233

Artificial system, 80, 83, 87-88, 98
Atavism, 118, 171, 177-178, 184, 207

Bacon, Francis, 240
Baer, K. E. von, 121, 122
Barlow, Lady N., 34
Barnacles: monograph on, 10, 78, 104, 195; classification of, 85, 98, 104, 126; Darwin's reasons for studying, 103-105, 111, 117, 128-129; relation to Darwin's other work, 104-105, 112-118 *passim*, 124, 128-129, 134, 144
Beagle, H. M. S., 2, 9, 10, 13, 31-34 *passim*, 104
Bees. *See* Social insects
Behavior. *See* Psychology
Behaviorism, 190-193, 209
Bell, C., 205
Belon, P., 106
Bentham, G., 32
Biogenetic Law. *See* Recapitulation
Biogeography, 32-43 *passim*, 48, 99, 100, 129, 236
Biological species: defined, 53, 55-56, 90-91, 192; and Darwin's concept, 101-102
Blending inheritance, 66, 133, 140, 160-180 *passim*
Blyth, E., 47
Bridgewater Treatises, 136
Brown-Séquard, C. E., 182
Buffon, G. L. de, 47, 53, 148
Butler, Samuel, 46-47

Cahn, T., 125
Catastrophism, 14-15, 251. *See also* Methodology: Uniformitarianism
Categorical rank, 84, 95, 97, 99

281